MATHIAS BERNHARD
DANA CUPKOVA
FUJAN FAHMI
CHRISTOPHE GIROT
FABIO GRAMAZIO
FRANCESCA HUGHES
ILMAR HURKXKENS
MARCO HUTTER
DOMINIC JUD
MATTHIAS KOHLER
BENEDIKT KOWALEWSKI
JESÚS MEDINA
AMMAR MIRJAN

ROBOTIC LANDSCAPES—
DESIGNING THE UNFINISHED

REFLECTION

DIGITAL MATERIALITY IN LANDSCAPE ARCHITECTURE
16–21

TOWARDS BIO-TECHNOLOGICAL INDETERMINACY:
OR HOW TO SHOVEL WITH PLEASURE TO CARE FOR THE WORLD...
22–33

THE STATE OF PLAY:
DIGGING IN THE SANDBOX OF ALL THE WORLDS
34–45

THE STORM, THE ARCHITECT, AND THE ROBOT:
LANDSCAPE MAINTENANCE IN THE AGE OF ROBOTIC INTELLIGENCE
46–59

OPERATION

SURVEYING
106–119

PROTOTYPING
120–131

PROCESSING
132–149

MANIPULATING
150–163

SIMULATING
164–177

EVOLVING
178–191

MAINTAINING
192–207

FOUNDATION

AUTONOMOUS EARTHMOVING
64–69

MODELING THE FIELD
70–75

SIMULATING MATERIAL FORMATION
76–81

THE ROBOTIC SKETCHBOX
82–87

TOPOGRAPHIC INTRICACIES
88–101

INTRODUCTION

No work of art remains immune to the effects of time. In terms of cultural connotations, a physical object or structure not affected by the passage of time is referred to as timeless.[1] In the life of an object, it comes closest to perfection at the moment of creation, but with the hope that its appreciation will endure well beyond this point. An object's fabrication, content, and materiality are inherently informed by the beliefs and production processes of the cultural epoch in which it is made. As such, even if it is conceived to be everlasting, its symbolic dimensions and physical appearance can never be timeless: a work of art will always be marked by the trace of time. While one can distinguish the longevity of its underlying structure compared to its surface, it is clearly a receptacle for, and the result of, the forces of nature.[2]

If we rate the success of a work in the built environment on its invariable qualities, then any shift from its initial form would seriously compromise its value or function. Nonetheless, built landscape designs are delivered as completed works and it remains somehow unimaginable that a project would evolve beyond its initial form. *Robotic Landscapes—Designing the Unfinished* proposes that we approach the processes of conceiving and constructing independently from such formal constraints and embrace all the temporal processes that naturally affect these structures. In this practice, the designer pursues a strategy of foresight and prepares the project for a continuous, dynamic evolution one capable of continuous adaptation to a rapidly changing environment. Instead of formally fixing a project in what will inevitably engender its slow decay, depreciation, and dysfunction, a dynamic project could sustain its purpose and meaning over time.

The notion of resilience can be thought of as a project's ability to withstand difficult conditions or to recover quickly by returning to its original form. However, progressive practices are moving away from the pursuit of permanence towards a more open, immanent, collective, and creative approach to landscape design that focuses on performance. The design strategies presented here explore how new environmental realities can be integrated into existing conditions, how designing landscapes to allow for change over long time periods may reap significant rewards, how the temporal can be exploited to our advantage, and how working with found materials promotes sustainable and ecological systems. Ultimately, experiencing the passage of time enhances our connection to the natural world. As such, it is a fundamental strength to constantly let an original proposal evolve, from the earliest sketches and models, through its construction, and beyond the maintenance work that follows. The design chain is therefore based on all phases of a project, with every detail in the original concept rooted in designing procedures that evolve with the changing landscape structure over time. Instead of fixing a static, immutable object from the onset, this book calls for a dynamic and open landscape approach that actively interacts with the natural flow of materials.

Robotic Landscapes compiles a series of ground-breaking experiments conducted as an interdisciplinary collaboration between landscape architecture, environmental technology, and robotics, exploring the interrelationships between varied granular materials, design procedures, computational logic, and natural processes in design. Through multiple time frames, it further investigates the transformation of locally sourced materials into operative landscape structures at various scales. Reconciling landscape forms and processes, a dynamic constructive approach is born that recognizes the power of natural forces as performative and meaningful agents of architectural landscape design. We illustrate how landscape formations generated by robotic processes can be conceived to respond to unpredictable futures, where performative terrain structures that emerge from computational procedures will become capable of sustaining a dynamic landscape system in continual flux. As such, this book pushes the concept of the unfinished landscape to new limits, proposing new views and methods that address many challenges found in the built environment today. Accepting and embracing the continual metamorphosis of terrain means landscape architecture no longer produces static projects fixed in a single moment in time, but rather that it works on specific terrain formations that are influenced knowingly by timely robotic interventions that learn to interact with the inevitable progression and transformation of the natural world.

Designing dynamic landscape construction processes within an unstructured environment poses serious challenges in terms of methods and medium, because a site is always characterized by unpredictable change and natural transformation. In order to respond to real-world challenges through the ongoing transformation of landforms, a series of design and construction experiments developed at ETH Zurich have given rise to new methods linking three distinct phases of the landscape design process. Successive design schemes are generated through an iterative loop comprised of actions in framing, forming, and finding. The contemporary landscapes shown in the figures of this book propose landforms that perform as successive events that guide and support the processes in which they are embedded. Accepting the entropic nature of our world, resurgent qualities may generate a better understanding between that which can be found on site and that which can be built. We deem this paradigm shift as essential to any resilient landscape practice today. "Designing the unfinished" must become inherent to the actual acts of design, construction, and continual landscape transformation.

Given that landscapes are perpetually transforming, what tools and processes can be utilized to establish a dynamic design and construction approach capable of responding to continuous change? The rise of computational methods has marked a shift away from a static and finite understanding of design towards a more active acknowledgment of formation processes. For landscape architecture, computational tools can help with the simulation of evolutionary processes in nature. Through the establishment of a digital design chain, they can guide adaptive processes in robotic fabrication that are essential to a dynamic understanding of terrain. Through digital technologies, it has now become possible to mediate erosion, transportation, and sedimentation over time in an ever-changing environment, as opposed to creating predefined and static earthworks. Computation opens an entirely new way of conceiving and designing landscapes, specifically the ongoing natural transformation of a territory. Instead of designing the actual form of a landscape, new procedural terrains emerge that negotiate natural processes through a robotic response. Dynamic landforms thus arise from complex relations between landscape forms and processes, enabling site-specific, adaptive, and emergent earthmoving scenarios that remain open-ended.

Due to the novelty of this approach, *Robotic Landscapes* opens with four essays outlining the future potential of the built environment. These essays open up a broader discourse on the current state of computational thinking in the landscape architecture discipline and bring insights into possible future developments. Five articles then explore the methodologies that establish the foundation of the specific design operations, leading to the integration of computation and robotic fabrication within landscape design. A selection of experiments illustrates the importance of a continual computational feedback loop between framing, forming, and finding, thereby facilitating the conception of a landscape through a better understanding of its present status. A compendium of images then carefully unfolds and reveals an entirely new kind of aesthetic, designed according to these computational principles. Organized into seven themes that accompany the process, each illustrates a selection of design research experiments on varying sites facing strong environmental challenges.

Translating the workflow of a dynamic design process into a printed publication has been a challenge. Instead of presenting iconic images, we chose to couple outcomes with early design sketches to highlight the process in a corresponding visual language. This also explains how the book was conceived, reflecting the production of the landscape as a printed sequence of steps that create meaningful superimpositions. As such, the unbounded nature of the design process was explored on a conceptual and a formal level of representation. In a seemingly unending sequence of frames, new and radically different design approaches—aiming at a new methodology, aesthetic code, and way of thinking—are choreographed throughout the book. The term "designing the unfinished" emerged as a guiding principle in the transformation of these new landscapes, resulting from complex computational processes that constantly shift their response according to environmental conditions and thus remain perpetually unfinished.

Robotic Landscapes tests a new kind of approach where the physical specificity of a site is integrated into the process of design and fabrication from the onset. The original form of a terrain acts as an initial computational catalyst and ordering principle that evolves in different scales in space and time. Responding to existing landscape structures is important in identifying the ways in which many future ecologies could be supported. We understand terrain as a load-bearing structure in the landscape and we aim to promote its aesthetic dimensions through computational response and performance. The intricate topographies shown in this book highlight the topological specificity of a given site and through constant progression one will never experience the same place twice. Dynamic planning and execution instruments open previously unthinkable design possibilities in the making of landscapes in the 21st century. It is essential to situate ourselves within the entropic processes inherent to a landscape by maintaining and actively developing forms that adapt best to unpredictable events. Acknowledging this continuous evolution not only gives value to the beauty of a site, it provides increased resilience as well. A renewed understanding of the unfinished helps us reimagine our environment as a truly timeless project—one in continuous transformation, shaped by the materials and processes that constitute it, as well as by the societal beliefs and desires by which it is conditioned.

1. In 15 BCE Vitruvius suggested that the timeless properties of architecture can be ascribed to *firmitas* (durability), *utilitas* (functionality), and *venustas* (beauty). See Vitruvius Pollio, *Vitruvius: The Ten Books of Architecture*, trans. by M. H. Morgan, Cambridge: Harvard University Press, 1914.
2. In this publication we refer to works of art in a broader sense, specifically to earthworks serving a certain large-scale task or activity. The German *werk* as well as the Dutch *kunstwerk* might be more precise words for the theme of this book as they are also used to describe large-scale hydraulic or geotechnical structures beyond the activities of construction or repair. See "Werk, n.4" in Duden Wörterbuch, https://www.duden.de/rechtschreibung/Werk (accessed May 15, 2021) and "Kunstwerk, n.2" in Van Dale Woordenboek, https://www.vandale.nl/gratis-woordenboek/nederlands/betekenis/kunstwerk (accessed May 15, 2021).

REFLECTION

What on earth do robots have to do with landscape design? Recent developments in computational tools and robotic fabrication have helped accelerate a new kind of approach to terrain. This shift is grounded in the belief that digital technology can mediate increasing public concerns about rapidly deteriorating ecological systems. In an era of increasingly complex environmental challenges, a focus on the way granular materials perform in a landscape seems to hold potential. Here, the primary emphasis lies not on form, but rather on the principles of the physical transformation of natural sedimentary matter. Digital procedures can adapt to several scales of territory and time frames, creating a new design approach that integrates environmental forces, societal concerns, and autonomous construction technology. This new form of methodological triangulation is explored in the essay *The Storm, the Architect, and the Robot* by Christophe Girot and Ilmar Hurkxkens, explaining the occurrence of natural material processes on terrains and the digital robotic processes that can react to them in an inherently dynamic way.

Digital Materiality in Landscape Architecture by Fabio Gramazio, Matthias Kohler, and Ammar Mirjan expands on the potential of robotic agency in landscape construction. This text reveals how robotic sensors establish feedback loops to address the dynamic behavior of granular material in complex, evolving geometries. Open-ended works could drive the transformation of locally sourced material into highly complex landscape formations, creating previously unthinkable design possibilities. As a noteworthy outcome, this suggests that autonomous excavating machines may soon become fully fledged actors in the landscape.

When earthmoving robots actively participate in the future shaping and evolution of our landscape, one question remains—for whom will they shovel? *Towards Bio-Technological Indeterminacy* by Dana Cupkova critically accesses environmental design procedures between mechanistic and substantialist worldviews. The essay propounds the inclusion of all living systems as drivers in future landscape design, refraining from a focus on humans on one side and ecology on the other. A more emphatic, all-inclusive approach to landscape design could potentially benefit and reinforce our connectedness to a one-sided natural world and, in turn, create an entirely new relation to technology.

If we accept the fact that ecology has its own agency and rights, then robots should be able to claim this right as well. In *The State of Play*, Francesca Hughes questions this technological optimism based on a robot's practical abilities, and argues for more indeterminate and creative ways of production. Are chaotic, sometimes even catastrophic natural events not exactly why we remain in awe of the natural world? Alluding to the limits of optimization and control—or even the futile pursuit of precision—Hughes outlines why there might actually be smarter ways of working as we computationally engage and interact with unstructured dynamic environments. As we strive for purpose, does a robot not just want to play?

DIGITAL MATERIALITY IN LANDSCAPE ARCHITECTURE

FABIO GRAMAZIO
MATTHIAS KOHLER
AMMAR MIRJAN

1. For a more in-depth description of HEAP, see Dominic Jud, Marco Hutter, "Autonomous Earthmoving," later in this book.

THE ADVENT OF SENSORY CONSTRUCTION MACHINES Technological developments in sensing, computation, and control have led to the creation of new robotic systems by digitally upgrading established construction machines. Such robotic construction machines have profoundly different capabilities from their human-operated predecessors; they are more versatile, agile, and—similar to industrial robots—they can autonomously execute a digital design blueprint without human intervention. An example of this new kind of construction machines is the robotized walking excavator developed at ETH Zurich.[1]

In contrast to conventional digital fabrication machines such as industrial robotic arms or CNC-routers installed at fixed positions in a factory environment, robotic construction machines need to be mobile in order to be able to build on site. They depend on the real-time gathering of information about their environment to locate their own position. Thanks to the processing of visual sensor information, the machine becomes aware of its context as well as of its own spatial configuration in relation to it. Therefore, one can understand such a machine as not being detached from its surroundings, like their robotic predecessors, but embedded in it. One might even say that active sensing makes such a robotic machine become part of its environment. This real-time visual feedback also enables robotic construction machines to continuously measure, compare, and adapt to changes in their surroundings. Most importantly, these machines are not only aware of their own spatial configuration and the environment at a distance, but also of the forces acting on their joints and construction tools while physically interacting with material. The machine therefore directly engages with the physical materiality and the dynamics of its surroundings.

What does the advent of such sensory construction machines mean for the design of landscape architecture, and will it lead to a change in design practice? Will we witness entirely new design processes like those introduced to architecture with the advent of digital fabrication?

Digitally controlled machines that are aware of their surroundings—and have the ability to move, walk, or climb complex terrain in the case of the robotic walking excavator—dramatically extend their operational range compared to conventional, stationary digital fabrication equipment. As a result, untethered and autonomously moving robotic excavators extend the reach of computational design and digital fabrication methods to the scale of landscape architecture, profiting from the advancements of its sister discipline, architecture. This said, it might in turn be that landscape architecture becomes a significant driver of these developments in the future due to its clear dependencies on sensory construction machines to manipulate terrain, but equally through its open attitude to process-oriented design.

2. *Olzweg* consisted of a labyrinth of recycled glass sticks which appeared to be glued to the existing building and its courtyard. The labyrinth was constructed and rearranged by a robotic arm and randomized interactive software that continuously reprogrammed its parameters during construction.

OPEN-ENDED LANDSCAPE FORMATIONS Taking robotic landscape construction forward, one can imagine that digital design will enable several different excavators to share the same physical and virtual space. The robots will have complementary abilities, tool sizes, and payload capabilities, and collaborate on complex landscape construction and transformation tasks.

Such a choreographed formation could follow a predetermined digital blueprint. However, it could equally be designed as an open-ended process where design intentions do not need to define a target geometry but can be thought of as designed procedures that manipulate landscapes in exchange with, and reacting to, natural formation processes. In 2006, architecture studio R&Sie(n) put forward the prospect of an "unachieved process of construction." In the speculative project *Olzweg* at the FRAC Centre in France, an oversized robotic arm built a self-programming and progressively adapting labyrinth over the course of ten years.[2] In this narrative, the machine does not just build the structure but becomes an indissoluble part of it. Similarly, autonomously operated landscape construction machines challenge the established relationship of construction and maintenance. Rather than limiting their deployment to initial implementation, robotic landscape design systems could bring forward the concept of permanent constructions in which, as with robotic lawn mowers, the machine becomes part of the design of the environment. Conceptualized as an open-ended to process transforming the continuously eroding and decaying natural environment, robotic landscape design systems can incorporate autonomous machines that dynamically respond to change and adapt to novel and unpredictable situations. This requires an equally open-ended, soft computational design approach in which the model learns from failure and thus continuously evolves.

TOWARDS LARGE-SCALE DIGITAL MATERIALITY Zooming in on the physical act of robotic construction itself, digital control enables the synchronization of the numerous axes of the excavator into simple motion instructions. As designers, we can therefore directly exploit these capabilities and translate, for example, a spline-curve drawn in CAD software into a complex three-dimensional motion that directly interacts with the terrain. Given that a robotic excavator has enormous force, it can carve this line into the ground with the chosen tool as well as lift, orient, and place objects that weigh several tons with centimeter precision. As such, robotic construction machines expand the current scope of digital fabrication beyond human abilities. Through visual and tactile sensing, such a machine can precisely control complex material manipulations that involve impressive forces. It can, for example, adapt to the softness and hardness of inhomogeneous materials, "sensing" with its excavator shovel like a human senses with a finger. In turn, this directly expands the scope and reach of the designers who define what the machines should sense, react upon, and do in a particular context.

3. Fabio Gramazio, Matthias Kohler, *Digital Materiality in Architecture*, Zurich: Lars Müller Publishers, 2008.
4. Fabio Gramazio, Matthias Kohler, Jan Willmann, *The Robotic Touch: How Robots Change Architecture*, Zurich: Park Books, 2014.

The landscape architect can thus directly implement the procedural rules of the design execution with all the spatial, functional, and aesthetic consequences it entails. This approach leads to a unified understanding of digital and physical processes—of data and material—where the inclusion of feedback processes and material behavior relativize the primacy of static geometric form in the design of landscape architecture. We observed that this synthesis leads to what we refer to as Digital Materiality,[3] where computational design and digital fabrication methods are present in a novel material expression of architectural design.[4] We can therefore speculate about Digital Materiality at the scale of landscape architecture, where robotic fabrication—either with already known or novel construction processes—will shape the environment by including both natural processes and human agency. Digital Materiality can express itself with classical construction processes, but it can also do so with materials and vegetation that are locally sourced, assessed, and redistributed on site. We can begin to imagine highly articulated landscape formations, terrains with unprecedented degrees of openness, and uncontrolled naturally formed landscapes that have been neither seen nor experienced, literally creating new ground for future ecologies.

DANA CUPKOVA

**TOWARDS BIO-TECHNOLOGICAL
INDETERMINACY:**
*OR HOW TO SHOVEL WITH
PLEASURE TO CARE
FOR THE WORLD...*

"The fact that a cloud from a minor volcanic eruption in Iceland—a small disturbance in the complex mechanism of life on the Earth— can bring to a standstill the aerial traffic over an entire continent is a reminder of how, with all its power to transform nature, humankind remains just another species on the planet Earth."
SLAVOJ ŽIŽEK [1]

"Do you shovel to survive, or survive to shovel?"
KŌBŌ ABE, THE WOMAN IN THE DUNES [2]

The cycle of nature's violence has always been a part of human history. The era of the Anthropocene acutely accentuates the ecological imbalance between the mythological harmony of nature and its evolving brutality. As of 2020, humanity officially became the maker of the planet: according to research published in *Scientific American*, synthetic objects made by humans outweigh the living biomass of the planet, with a radical decline of plant biomass specifically.[3] "Just another species," as Žižek likes to refer to humankind, that created an irreversible footprint in its amassing of new matter. Architecture, landscape architecture, and the construction industry are complicit agents in the formation of this ossified garden in which waste streams became the new anthropogenic naturality, and the only future source of making environments anew. The scale has tipped irreversibly.

1. Slavoj Žižek, "'O Earth, Pale Mother!'," in In These Times, https://inthesetimes.com/article/o-earth-pale-mother/ (accessed June 17, 2010).
2. Kōbō Abe, *The Woman in the Dunes*, trans. by E. Dale Saunders, New York: Alfred A. Knopf, 1964.
3. Stephanie Pappas, "Human-Made Stuff Now Outweighs All Life on Earth," in *Scientific American*, https://www.scientificamerican.com/article/human-made-stuff-now-outweighs-all-life-on-earth/ (accessed Dec. 8, 2020).

This reality mainly comes from the deterministic, solution-driven approach of materialist design thinking that emerged from the industrial era—an inherent product of our design and production tools. Advanced computing, AI, and robotic construction processes lend a new urgency to the discipline of landscape architecture. This is primarily in terms of nimbly rethinking models of environmental transformation, and new design practices that promote architecture and urban form to be more active and reciprocal agents in rebalancing (or reshoveling) the structure of the built environment. Conceiving of architecture as a living system, as a part of large-scale human ecology, forces us to reframe the role of industrial infrastructure towards a bio-synthetic model. Landscape has a critical role in choreographing this interchange—it ceases to be a static plane with a determinate form. It becomes a dynamic force that carries and participates in the redistribution of time, matter, information, and species; a *deep* substrate allowing us to play out design choreographies in which the natural patterns of ecological emergence shift and coalesce over time. The issues of landscape depth and its representation are important as they allow for consideration of more complex relational "actants" to engage in formations of pattern ecologies;[4] to go beyond the superimposition of image-based surface features, transposed through contemporary texture modeling techniques so prevalent within the current AI style-transfer-based design experiments. Actants—all the living or nonliving things that

4. Jane Bennett, *Vibrant Matter: A Political Ecology of Things*, Durham: Duke University Press, 2010, p.viii.

constitute the politics of ecology—have rights by default, according to Jane Bennett, and need representation as they cannot speak for themselves. Yet bio-synthetic design frameworks combined with robotic agency propose to go beyond the largely insufficient problems of disciplinary representation. It allows us to design and perhaps re-shovel the temporal reality of our ecological emergencies in situ.

The Robotic Landscapes RL design research studio at ETH Zurich proposes a bio-technological design approach in its attempt to engage with this issue of landscape depth. The studio's experimentation is focused on site-specific robotic manipulation of granular matter, collapsing representation and real-time data to curate formations of new terrains. This approach attempts to rebalance the equilibria between social systems effected by natural disasters and infrastructural necessities of dynamic landscape transformation over time, positioning robots as caretakers of the land. Not unlike Kōbō Abe's novel *The Woman in the Dunes,* the studio acknowledges the immediacy of environmental disaster—of a continuously sinking ground overlapping with an aesthetic project that combines the removal and recombination of shifting instability into new forms of survival.

The designer in the RL design research studio becomes part of an orchestra that conducts human desires, choreographs natural elements, shifts granular matter, sequences geological layers, automates machines, and produces code. The design hierarchy is redistributed, and it remains in flux. The issue of control is ambiguous; at the forefront is the immediacy of responsiveness, questioning the predetermined design bias through the curation of a data set. This is not to claim some form of unbiased innocence, as organizing data structures already carries a bias of design intent, but this setup allows for a reactive immediacy that encourages us to learn and adapt over time. Just as ecological boundaries are fuzzy and indeterminate, these new patterns of landscape formations are conceived as shifting in time with varied input and feedback from material experiments and simulations, thus collapsing the space between analysis and design. Resulting representations are less focused on determining specific figurations and more on describing the instantiation of terrain shapes, as they materialize, into parts of industrial ecology. Immersed in a study of indeterminacy, they rely on the naturalizing behavior of robotic technology to stabilize emerging formations.

Expanding on the experimentation in formal approaches to urban ecology, the RL design research studio adds, as the name suggests, robots—in this case participatory caretaker robots engaged in the periodic and dynamic maintenance of landscapes. Moving away from a predetermined solution space and the static effects of an image, the studio's drawings emerge in the form of guidelines codifying temporal transformations in response to large-scale environmental forces. These include mudslides, avalanches, and floods, and are rooted in a combination of data and ecological patterns of climate change, as well as in proposed new patterns of human-landscape cohabitation without a deterministic divide between the constructed and the natural. These new terrain formations are both socio-ecologically operational and caringly beautiful. The RL design research studio proposes a design practice in the form of a high-stakes caretaking that represents the retooling of work and pleasure in a time of ecological crisis. In this case, we do "shovel to survive," but we also "survive to shovel" with pleasure (and with robots).[5]

The question of ecological projects in architecture and design is relatively new, lacking a more comprehensive historical dimension beyond the last three decades. Although landscape ecology has been at the center of landscape design since its introduction by Richard T. T. Forman in the 1980s, it was primarily situated in environmentalism turned into a quantifiable scientific metric that eroded any aesthetic or socio-cultural design project.[6] The advancement of simulation and computational design capabilities enabled the emergence of new landscape-integrated design strategies and nondeterministic approaches to time-based, process-oriented design embraced in the last few years.[7] Framed primarily in response to biological research on decreasing biodiversity, a biophilic integration of living systems into design thinking has taken many formal approaches including dimensions of sedimentation strata and their social implications.[8] The connection between landscape and

---

5. See note 2.
6. Richard T. T. Forman, *Landscape Ecology*, New York: John Wiley & Sons, Inc., 1986.
7. Bradley Cantrell, Zihao Zhang, "Choreographing Intelligent Agents," in BLACK BOX: Articulating Architecture's Core in the Post-Digital Era, 107th ACSA Annual Meeting Proceedings, Pittsburgh, 2019, https://doi.org/10.35483/ACSA.AM.107.1 (accessed Jan. 1, 2021).
8. See, for example, Scape Studio, "Public Sediment: Resilient By Design Challenge," in https://www.scapestudio.com/projects/public-sediment-resilient-design-challenge/ (accessed May 1, 2020).

architecture is most widely centered on the concept of biophilia, a term coined by Erich Fromm in 1973 and later popularized by E. O. Wilson in 1984,[9] arguing for the innate human need to connect with nature, and "the [human] urge to affiliate with other forms of life."[10] An aesthetic response to the architecture of biophilic form-making has emerged, mainly focused on the visual complexity of naturalism and material decay.

In most design practices and schools of architecture, the question of ecology is still being framed primarily through the lens of an ethical and/or sustainable imperative, dependent upon one's implicit or explicit attitude towards environmental ethics. In a theory on political ecology, Anna Bramwell places the rise of ecologism in the 19th century between two distinctive strands of thinking.[11] The first was influenced by Ernst Haeckel's holistic, anti-mechanic approach to biology, and the second focused on the problem of scarce and nonrenewable resources, an approach defined by energy economics. This duality of approaches is rather rampant in the current academic landscape of ecologically concerned architectures, inherently setting up dialectics between substantialism and aesthetics in design, with a desire for a singular solution. Substantialism typically relies on metric-based approaches towards positive accounting of environmental *goodness* and participation, and the aesthetic approach explores new representations of social and cultural narratives in the role of nature in the architecture of the Anthropocene. This dialectic

9. Erich Fromm, *The Anatomy of Human Destructiveness*, New York: Henry Holt & Co., 1992.
10. E. O. Wilson, *Biophilia*, Cambridge: Harvard University Press, 1984.
11. Anna Bramwell, *Ecology in the 20th Century: A History*, New Haven: Yale University Press, 1989.

fallacy of ecology as a mechanistic reality, and ecology as a substantialist reality, has been rooted in anthropocentric solution-space, and was first questioned by Howard T. Odum's early framing of systems ecology[12] as a "thermodynamic reality."[13]

Furthering the anthropocentric dialectic in a more nuanced way, Richard Ingersoll forefronts the paradox of natural and constructed environments, writing that "every act of building betrays the environment."[14] Focused on examining varied approaches to naturalism in the design of 20th-century architecture, Ingersoll describes categories of design imperatives through the lens of humanist and positivist thinking that range from "building against nature" to the "green apocalypse," and the "advent of land architecture" to "ecological correctness" and "quantification." Ultimately, this review of design's approaches to the environment ends with an open-ended question about future worlds, advocating alternative design practices connected to philosophical foundations in policymaking, catalyzing new frameworks for the best practices. At its conclusion, it avoids the question of human sensibility and assumes the necessity, or burden, of the ecological within constructed environments.

Concepts of the "political ecology of things" that cross over to new forms of materialism, such as Bennett's writing on "Vibrant Matter," are inspirational and provide an interesting framework for better understating the bio-

12. Howard T. Odum introduced the concept of "emergy" as a measure of qualitative differences between forms of energy, the diagrammatic mappings of which are based on Alfred J. Lodka's predator-prey diagrams. See Howard T. Odum, *Environmental Accounting: Emergy and Environmental Decision Making*, New York: John Wiley & Sons, Inc., 1995.
13. For more about Odum's ecological system framework, see Dana Cupkova, "Eco-machinic Others: From Polarities to Gradients," in *The Expanding Periphery and the Migrating Center*, 103rd ACSA Annual Meeting Proceedings, Toronto, 2015, pp.594–602.
14. Richard Ingersoll, "The Ecology Question and Architecture," in *The SAGE Handbook of Architectural Theory*, Washington: Sage Publications, 2012.

technological approach.[15] Bennett's writing proposes a new hierarchy in which a construction of public interest includes all living systems—worms and humans alike. This type of "public" escapes the imitation of traditionally defined participatory design and limited forms of social collectivism that often fall into a top-down model of community saviorism, an effort biased by the "informed" and "engaged" designer. Bennett's definition potentially avoids the pitfall of good intentionality that landed idealist modernist architecture in the hands of totalitarian systems. In Bennett's proposition, the construction of public shifts away from Marxist materialism, erasing gaps between agencies of matter and human and nonhuman actants. Matter is not a product, a resource, or a commodity, but part of a waste stream life cycle, a living entity, "vibrant matter" in its own right.

This shift from classical materialism towards the "vibrancy" of matter suggests another level of connectedness, or the agency to connect geological and living matter with a more differentiated definition of public into forms of care.[16] If the worm is an actant in the life cycle of any material construction, it has a role within the hierarchy of political decision-making that could aid in large-scale territorial design processes. If design could be grounded in an environmental empathy that leads to more diverse paradigms in the redistribution of resources, it could position architecture as a vehicle for ecological and communal restoration. Centered on cycles of landscape waste

15. See note 4.
16. See Donna Haraway's definition of a difference: "My hope has been that the always oblique and sometimes perverse focusing would facilitate revisionings of fundamental, persistent western narratives about difference, especially racial and sexual difference; about reproduction, especially in terms of the multiplicities of generators and offspring; and about survival, especially about survival imagined in the boundary conditions of both the origins and ends of history, as told within western traditions of that complex genre." Donna Haraway, *Primate Visions*, New York: Routledge, 1989, p.377.

streams and material-embodied energy, we could discover new forms of co-shared domesticity, as well as social equity within our collective urban space, closely entangled within its ecological functions. The question is: can we give robots enough agency to care for worms and allow for these forms of collective design empathy?

This concept of empathy and "connectedness," as opposed to the "truthiness of a data dump" discussed by Timothy Morton, frames ecological knowledge as a form of "attunement."[17] "Since a thing cannot be known directly or totally, one can only attune to it, with greater or lesser degrees of intimacy. Nor is this attunement a 'merely' aesthetic approach to a basically blank extensional substance. Since appearance cannot be peeled decisively from the reality of a thing, attunement is a living, dynamic relation with another being."[18] Thus, ecology in architecture raises a question of its representation, of worms, of humans, of nonhumans, to present inherent otherness to the familiar.

The approach of the RL design research studio carries a fascination with both the beauty and violence of landscapes, their inherent familiarity and simultaneous otherness. This design process attempts to avoid the obliviousness by which we tend to situate objects within a territory of representational stasis. Here, the landscape does not slow down; it maintains its dynamic nature, both operationally and representationally. It does not stop for a moment to receive an architectural object.

17. Timothy Morton, *Being Ecological*, Cambridge: MIT Press, 2018.
18. Timothy Morton, *Ecology Without Nature*, Cambridge: Harvard University Press, 2009.

This daring choreography of operations is rooted in topological geometries that cross-reference robotic computation with the natural tendencies of accumulated matter to form a temporary stability. As our human and geological times collapse more closely due to effects of climate change, the natural hazards are more acutely prevalent, building upon each other. Here, they are managed by robotic routines that allow new patterns to emerge on top of the old, without the effort of large-scale removal and repair typically required by industrial infrastructure repairs. As robots slowly care for, and shovel, shifting layers of soil to periodically restabilize terrains, they tame the potentials of new landscape violence. Distributed landscape patterns are part of a formal response in which the waste streams of debris participate in a computational model and are tested to produce new stabilized configurations. A force of anthropogenic nature facing the power of robotic pseudo-intelligent agents producing new temporary forms of picturesque debris. These landscapes of mudslides and moving earth are part of accumulative indeterminate ecological patterns that occur in real time through periodic robotic participation. Intuitively, this process reflects on a desire for a new form of collectivity. "If we stop shoveling," the woman in the dunes explains, "the house will get buried. If we get buried, the house next door is in danger."[19]

The work of the Robotic Landscapes design research studio positions landscape as a substrate for both scientific inquiry and the exploration of an

19. *Suna no Onna (Woman in the Dunes)*, dir. by Hiroshi Teshigahara, Tokyo: Toho Film (Eiga) Co. Ltd., Teshigahara Productions, 1964. Scene mentioned in Roger Ebert review, "Woman in the Dunes," https://www.rogerebert.com/reviews/great-movie-woman-in-the-dunes-1964/ (accessed Feb. 1, 2020).

aesthetic language, putting pressure on predetermined design space. It promotes a shift away from the data-driven rationales of performative models based in a singular time and space that exhausts their breath and fidelity, largely by reducing context to a fixed information set. Queried through measured simulation sets in a singular moment in time and space, the studio explores a new form of real-time dynamism. Simulation is used to provide dynamic readings of a site's ecology, representations of which are cyclically linked to physical formations of objects in landscapes.

This framework brings bio-technological design thinking to the fore to negotiate differences between organic and inorganic, formal and performative, cultural and ecological, zones and boundaries, solids and gradients, proximities and tendencies, objects and fields. Using formal codification of geo-spatial data sets, it reinscribes invisible forces into formed constructs with the hope of shifting the focus of design away from the singular contextual "truth" of data, towards the constraints of indeterminate spatial configurations, gently cared for by the robotic force.

THE STATE OF PLAY:
DIGGING IN THE SANDBOX OF ALL THE WORLDS

FRANCESCA HUGHES

1. See Robert Smithson, "The Monuments of Passaic" in *Robert Smithson: The Collected Writings*, Berkley: University of California Press, 1996, pp.68–74.

In 1967 Robert Smithson, camera in hand, takes a trip to his childhood town of Passaic, New Jersey. There he names and photo-documents six "monuments" from the familiar objects of its semi-obsolete industrial landscape, each cut adrift from or barely tethered to their attendant infrastructure: giant pipes spewing water or pontoons snaking across mud flats to reach their forlorn derricks. Of the six *Monuments of Passaic, New Jersey* the last one stands out, *The Sand-Box Monument*—a children's sandbox from which half the sand has escaped onto the playground grass.[1] We know from the park bench in the background that it is intimate in scale, only three small children might be able to crouch and intently dig and press within its square frame. The other five photographs claim their monumentality via scale, endurance, and fixed form (they could be Roman), whereas the sandbox is declared monumental despite having none of these attributes. Instead, it presents its performative *scalelessness* (it is to all intents and purposes the same thing no matter what size it is—Smithson calls it a "desert") and, in its ever-changing form, presents the ultimate enduring and fixed formal condition: the second law of thermodynamics. While the other monuments testify to the anti-entropic endeavors of all industry, ever brave in the face of futility, the sandbox has no delusions. Like a mirror—the many mirrors Smithson installs in the recurring sand pile and other projects in his work—it simply reflects the state of play.

2. Indeed, further running will only make the gray more uniform. Ever cunning, Smithson adds, "Of course if we filmed such an experiment we could prove the reversibility of eternity by showing the film backwards, but then sooner or later the film itself would crumble or get lost and enter the state of irreversibility." This is representation staged to offer a false escape from entropy. See note 1.
3. See Peter Galison, "Epistemic Machines: Image and Logic," in *When is the Digital in Architecture?*, Montreal, Berlin: CCA, Sternberg Press, 2017. A Galton board, also known as a quincunx, is a glass-fronted box into whose top balls of lead are dropped. These fall through rows of pins, offset like the dots of the number five on a die, and pile up at the bottom in a heap that traces a "normal curve." First presented by Francis Galton in 1893 in a lecture to the Royal Institution of Great Britain, this instrument effectively secured the new authority of the normal curve.
4. Ian Hacking, *The Taming of Chance*, Cambridge: Cambridge University press, 1990.
5. Roger Caillois, *La Dissymétrie*, Paris: Gallimard, 1973.

Smithson famously harnesses the sandbox's inherent indexicality as an epistemic machine through which to explain entropy. A child runs a hundred times clockwise in a sandbox filled with black sand on one side, white on the other. The churned sand turns gray, but no amount of running counterclockwise will reverse the gray and once more separate black grains from white.[2] Like a Galton board, the sandbox in Smithson's parable combines properties of both traditions of seminal epistemic machines: those that are visual and that make a picture of what they are explaining (such as a cloud chamber), and those that are indexical and for which no single measure generates meaning, but a set of measures does (such as a Geiger counter).[3] Not only can we watch the arrival of the gray, were we to choose to, we could literally count, grain-shift by grain-shift, our decaying energy state. The grains of sand in the sandbox are like the lead pellets in Francis Galton's 1893 board (or balls in the Probability Machine as Charles and Ray Eames renamed it when they set it center stage in their 1961 *Mathematica* exhibition). Both draw a picture, and present a data set to be counted, of what they are explaining, entropy itself or, in the case of the Galton board, entropy mediated via the binomial or normal curve—the Victorians' great anti-entropic project to bring the empire to heel through symmetry.[4] Intentionally or not, this legacy is affirmed with Roger Caillois' own term for entropy, "La dissymétrie."[5] The reassuring figure of symmetry does not stand up to the test of time's arrow (another alias for entropy); putting

6.   Robert Smithson interviewed by Alison Sky, see note 1, p.301.
7.   Norbert Wiener, *The Human Use of Human Beings*, Cambridge: Da Capo Press, 1954, p.12.

Humpty Dumpty together again is not quite as easy as falling off a wall. The Victorian mechanistic worldview (essential to the building of an empire in the first place) assumed reversibility in all things, but "entropy," as Smithson points out, "contradicts the usual notion of a mechanistic world view. In other words, it's a condition that's irreversible."[6]

    Josiah Willard Gibbs explained entropy through probability. There was no longer just one world but many—one for each possible outcome. The balls in a Galton board could be simply understood to represent all the possible answers to a question, all the possible configurations of black and white sand grains in a sandbox, say, in all their possible worlds. The worlds with gray sandboxes cluster in the middle, the very, very much rarer worlds with more black grains to one side of the box, more white to the other, at the margins. In Gibbs' universe of infinite worlds, order is least probable and chaos most probable, and it thus follows that over time chaos increases as the universe decays from "a state of organization and differentiation in which distinctions and forms exist, to a state of chaos and sameness."[7] However, as Norbert Wiener explains, this does allow for temporary pockets of exception—"enclaves" bounded by space and time, within which this trend can be temporarily reversed: "But while the universe as a whole, if indeed there is a whole universe, tends to run down, there are local enclaves whose direction seems opposed to that of the universe at large and in which there is a limited and

8. See note 7.
9. Erwin Schrödinger, *What is Life*, Cambridge: Cambridge University Press, 1992, pp.21–22.
10. Both Norbert Wiener, see note 7, p.25. The emphasis on "temporary and local reversal" is my own.
11. "Sooner or later we shall die, and it is highly probable that the whole universe around us will die the heat death, in which the world shall be reduced to one vast temperature equilibrium in which nothing really new ever happens. There will be nothing left but a drab uniformity out of which we can expect only minor and insignificant local fluctuations." See note 7, p.31.

temporary tendency for organization to increase," adding, "life finds its home in some of these enclaves."[8]

It was this same entropy-reversing "life" that caused physicist Erwin Schrödinger to turn in 1944 to biology in search of his "negative entropy," and in particular to the chromosome with its uncanny ability to preserve information and resist formal decay. Tellingly, Schrödinger names the chromosome—the key protagonist of his gene action theory—"architect's plan and builder's craft in one," that is, both instruction and execution.[9] Unlike Schrödinger's, Wiener's "life" was of course always a euphemism for feedback—itself, incidentally, the downfall of Schrödinger's chromosome-architect with its causal linearity scripting all life from cradle to grave. And Wiener's feedback is not that of the organism but of the machine: "The control of a machine on the basis of its actual performance rather than its expected performance is known as feedback and involves sensory members which are actuated by motor members and perform the function of tell-tales or monitors—that is, of elements which indicate a performance." This feedback, like that of the organism, builds an enclave: "It is the function of these mechanisms to control the mechanical tendency toward disorganization; in other words, to produce a *temporary and local reversal* of the normal direction of entropy."[10] All the king's horses and all the king's men may, after all, be able to put Humpty Dumpty back together again, but just for a short spell—a "local fluctuation" within the inevitable descent into "drab uniformity."[11]

```
12. This confusion is conscious. Less conscious
perhaps is the metaphoric confusion at large
in the discourses surrounding robotic landscape
construction technology in which—at the interface
of digital computation and the real—the term
"computing" is also used to describe the phenomena
of form-finding that occurs in the physical,
material world in phrases such as "the material
computes" or "natural computation". Just as AI
thinking is being projected onto our understanding
of the brain in current neuroscience, here we do
not have anthropomorphism but a kind of "computo-
morphism" projected onto materials. Weiner—with his
early and very strategic confusion of animal and
machine—would be delighted.
13. Maxwell's demon first appeared on December 11,
1867 in a now seminal letter to P. G. Tait.—one
of physics' more exquisite explanatory models. See
C. G. Tait, The Life and Scientific Work of Peter
Guthrie Tait, Cambridge: Cambridge University Press,
1911, pp.213-214.
```

The sandbox, with its framed, uniformly discrete, granular contents, whose contours are written with uniform precision or resolution no matter which coordinates they cross, is always already indexical, always already poised for digital logics; indeed its analogue loyalties are tenuous at best. There is no representation of the sandbox as it is already the best representation of itself. At 1:1, grain-to-grain resolution, it rehearses Jorge Luis Borges' 1:1 map of the world as part of a 1:1 sandbox of the world fit for the anthropogenic age, or as part of a 1:1 point cloud model of the world. In the ensuing confusion of real and represented (incidentally the lifeblood of all play, to which we will come) at work in robotic landscape construction, the granular terrains of the river-beds and mountainsides are recast as giant Galton boards—statistical models with an infinite sample set of their own inexorable decay.[12] These mountain-sides-as-models, at the resolution of reality, confirm the adage that higher levels of resolution lead to higher levels of noise. The noise is deafening.

Enter the robot and its not-so-nimble "fingers" or end-effectors, a latter-day echo of James Clerk Maxwell's "neat-fingered," entropy-reversing "demons" that sit at the threshold between two chambers of warm gas. There they patiently sort hot molecules into one, cold into the other, so that the "hot system [chamber] has got hotter and the cold colder and yet no work has been done, only the intelligence of a very observant and neat-fingered being has been employed."[13] Whereas the natural terrain is infinitely articulate in its diary of

14. Kate Gannett Wells, chair of the Massachusetts Emergency and Hygiene Association. See Alexandra Lange, *The Design of Childhood: How the Material world Shapes Independent Kids*, London: Bloomsbury Publishing, 2018.

descent into Wiener's "state of chaos and sameness," the robot-digger has two words with which to respond: "cut" and "fill" (maybe three if we include the "spread-and-compress" action of grading). Within its sub-cycles of operation, the point in the point cloud and the pebble in the riverbed have entered into a contract as mutual aliases. However, the robot's clumsy fingers, vocabulary of two, and the constraints of its execution pathways present a precision bottleneck in this exchange, so narrow it is tragicomic. Like the nymph condemned to empty the beach of sand with nothing but a thimble, only infinite time can make sense of the task. Otherwise put, this task, by definition, is not one to be finished.

    The sandbox is also a supreme site of play. The first sand-bergs (or piles of sand) appeared in parks in Berlin in the 1850s, influenced by Friedrich Froebel's concept of the garden in the emerging kindergarten typology. By 1880 they had become an established component in the design of kindergartens and were introduced to the United States by physician Marie Elizabeth Zakrzewska in the form of sand gardens. Five years later, twenty-one had been built for children in Boston immigrant housing for whom the then chair of the Massachusetts Emergency and Hygiene Association declared, "Playing in the dirt is the royalty of childhood."[14] However, childhood was soon to be lent a purpose. In 1887, American child psychologist G. Stanley Hall wrote *The Story of a Sand-Pile*—an account in which a large pile of beach sand is

15. G. Stanley Hall, *The Story of a Sand Pile*, New York: E. L. Kellog & co., 1897, p.3.
16. See note 15, p.18.
17. Roger Caillois, *Man, Play and Games*, trans. by Meyer Barash, Chicago: University of Illinois Press.
18. Johan Huizinga, *Homo Ludens: A Study of the Play-Element in Culture*, London: Angelico Press, 2016.

deposited beside an idyllic summer cottage outside of Boston. Hall observes the effects of this pile on his subjects: a group of well-to-do boys on vacation, prone to laziness and fighting, but for whom "all other boyish interests gradually paled" once presented with the sand pile. All summer long they made "wells and tunnels; hills and roads like those in town; islands and capes and bays with imagined water; rough pictures drawn with sticks."[15] Hall notes how the boys, some as old as fifteen, cooperated with each other and were rarely idle as they solved, in their sand world, the problems of the "real" such as "industrial processes, institutions, and methods of administration and organization."[16]

All play, as defined by Caillois, is itself always a spatially and temporally bound enclave. It is also a legislative enclave—suspending all ordinary law and creating its own legislation—and make-believe, or at least at odds with reality. Play must also be voluntary; you cannot be told to play. Hence, the parental imperative to "go play" is oxymoronic and always fails. Caillois' typological analysis of play in the 1958 text *Les Jeux et Les Hommes* (translated as *Man, Play and Games*)[17] was a revisionary critique of Johan Huizinga's seminal 1938 book *Homo Ludens*.[18] However, there are three points Huizinga makes that are important to retain in considering both Hall's subjects and the robotic players or "robo-ludens" that are the subject of Robotic Landscapes in their sandbox-of-all-the-worlds. Firstly, the many attempts to assign a

19. See note 18, p.6.
20. See note 18.
21. Conversation with Ilmar Hurkxkens, January 2021.

function to play (social, biological, educational) all miss the point as they fail to address the profound aesthetic quality of all play. Secondly, all play (not unlike error or matter) is ultimately defined by what it is not—it lies outside of the logics "wisdom and folly, and equally out of truth and falsehood, good and evil."[19] Thirdly, "in play, there is [always] something *at* play which transcends the immediate needs of life and imparts meaning to the action."[20]

So what exactly is *at* play in the antics of these robo-ludens? Consider the three-way meeting of the end-effector, the point in the point cloud, and the pebble in the riverbed. Precision bottlenecks are the site of extraordinary error. Ilmar Hurkxkens recounts how it is the resolution errors in the sandbox that make the robots "come alive" and start to play, to engage with entropy differently.[21] Conversely, play attends precisely to the logics that flourish at precision bottlenecks: silly things that are rendered difficult to do, or perversely unoptimized endeavors. Thus comprised, play is able to harness "La Dissymétrie" for its own delight and, not unrelated, sabotage the real, but also the deterministic, the transactional, and the predicted. Play would push Humpty Dumpty back off the wall again for the pleasure of the catastrophe, for the puzzle of the repair.

There are two further criteria to Caillois' play: it is uncertain (and thus embraces error) and it is unproductive, "creating neither goods, nor wealth, nor new elements of any kind; and, except for the exchange of property

22. See note 17, p.10.
23. See note 18, p.11.
24. See note 1, p.12.

among the players, ending in a situation identical to that prevailing at the beginning of the game."[22] The aesthetic project of play to which Huizinga refers (and that Hall's analysis eclipses in "Industrial processes, institutions, and methods of administration and organization") is precisely this: to undermine the very logics that require purpose, profit, and end product. What is at play in all play is the interrogation of the hegemony of *purpose* and the teleology that attends to the finished state. No true play is ever "finished," only suspended until next resumed, including that of our robo-ludens here. Huizinga further notes that society is always more tolerant of the cheat than the spoilsport whom, by pulling out of a game, shatters the suspension of disbelief and "robs play of its illusion—a pregnant word which literally means 'in play,' from *inlusio*, *illudere*, or *inludere*."[23] We might ask what, then, is the illusion under which these robots, as executors, and their architects, as instructors, are laboring? Might the illusion at play in *our* thinking about their play be that of false purpose? In particular, purpose derived in terms of a problem that fuels the solutionism to which, in our tunnel of technological determinism (our unswaying belief in an instrumentalist premise whose indeterminacy is never fully declared), we increasingly turn to as a technological *raison d'être*. Smithson is clear on the problem with problems: "Problems are unnecessary because problems represent values that create the illusion of purpose."[24]

Like all the best games, engaging with such a manifest display of entropic complexity saves us from our delusions of techno-determinist grandeur. The egg and spoon race comes to mind. At the end of the day, all these robots *can* do is build an enclave of play; this could also be both their best gift and our best laboratory. It is perhaps not surprising that in landscape architecture, unlike architecture, we find the delirious calculatory power that computers so glibly lay at our feet not used to produce shiny drawings or the highly articulated—yet grossly reductive—carapaces of parametric optimization. Instead, this power is employed in modeling the inherent indeterminacy in our processes, and in those of our subjects, in order to make enclaves of ulterior purpose (or deferred purpose) and intelligently engage with indeterminacy, error, and more ludic economies of production. As architects and landscape architects, the solutionism to which we turn (in our moment of intellectual laziness) or insist upon (in our moments of insecurity) only paints us into a corner that will always inhibit a more truly precise, more truly intelligent engagement with the full complexity of the physical world. It goes without saying this is an inhibition we can ill afford in addressing how we redesign our relations to technology (and, yes, entropy) in order to meet new climatic regimes. The mask has slipped. Technological purpose is an alibi we must now drop if we are to genuinely ask of computation's encounter with chaos: *what game are you playing?*

CHRISTOPHE GIROT
ILMAR HURKXKENS

THE STORM, THE ARCHITECT,
AND THE ROBOT:
*LANDSCAPE MAINTENANCE
IN THE AGE OF
ROBOTIC INTELLIGENCE*

PROLOGUE With a mix of tradition and innovation, architectural teaching questions the role of history, which carries the fears, memories, aspirations, and experiences of past generations. Architecture evolves gradually, creating concrete visions and projects through a given window of opportunity; each project must then get to grips with this reality through a set of cultural codes that are interwoven with countless political and financial burdens. When a project is finally realized under challenging economic circumstances, it embodies a facet of that epochal reality. Yet, the last 5,000 years of design have shown that societies are generally greedy, energy-consumptive, short-sighted, unpredictable, and disruptive.

Our current age of anarchic social media, our erratic behavior towards ecology, increasing pandemics, and changing global economies escapes any predictable rule or order. In this context our own comfort zone comes into question, and the temptation to cling to a pre-existing lore of design in the landscape is indeed strong, often leading us toward what we already have and know. This highlights a certain weakness in our relationship to the world, where we tend to struggle, more often than not, with figments of our imagination in the face of accrued environmental and societal challenges. In a society affected by waves of lasting change, what should become the role of the architect? How can we best prepare the practitioners of landscape design to respond to the forces that they will face tomorrow?

This essay will not debate the stylistics of contemporary design practice *per se;* instead, it proposes a point of view on the evolution of the architect's engagement in a context of confrontation (not to say conflagration) between the forces of society, the broader environment, and robotic intelligence.

A MANIFESTO FOR A NEW ARCHITECTURAL EDUCATION To bring clarity to the overwhelming design challenges we are facing, the authors propose a simple triangulation where three often juxtaposed factors merge into one: the Storm, the Architect, and the Robot. The most challenging factor in this triangle is the Storm, with its reality of shifting ecological goalposts. The Storm represents all the environmental challenges that now face society. The Architect's most daunting task is to shift our focus on this reality by embracing new skill sets. We should train future generations of architects as enablers and comprehensive thinkers who will come to understand the physical principles of design across several scales and disciplines. The definition of a new design approach to landscape will become a dynamic one, combining societal and environmental factors with artificial intelligence. Recent developments in computational tools and automated intelligence mean that the Robot is gradually becoming part of current architectural curricula. While the content and goals of computational learning are in need of better calibration and focus, the complexity of robotics and its adaptation to landscape projects has at last become demystified and manageable.

THE STORM The environment is as unpredictable as it can be violent and destructive. It speaks to us of a universal law of nature and, more precisely, the third law of thermodynamics: entropy. This constant in nature is a transformative driver that tends towards zero over the universal immensity of time, matter, and space. Brought back to the scale of Mother Earth, these are the forces of changing times that fret with the weight of matter over mind. Meanwhile, advanced stochastic calculations enable us to reach out and look at our world ahead through the tiny pinhole of science.

Understanding precisely what is coming at us—economically and culturally—in terms of climate change is a design challenge of another magnitude altogether. Knowing about changes to come and analyzing them does not necessarily mean that we can act upon or prevent them; therein lies the true dilemma of the contemporary architect. The pre-Socratic Greeks called their environment *physis*, signifying everything material and physical that could materialize in space and time outside the human body and society. For that reason, the Greeks did not mistake *physis* for a god, but instead understood the physical manifestation of our world as a given.

However, the scientific apparatus that we presently trust has mutated into a rather dogmatic mirror of the environment, one that translates nature into a political construct, fueled by excessive guilt, belief, and wonder. The laudatory posture and intrinsic fascination we show for nature also hides an overwhelming and ugly truth, that of an imminent cataclysm foretold. The environment today, as almighty and overwhelming as it may seem, has in fact become the threatening and wrathful God of our age, a Gaia that is both feared and adored, deadly and admirable. Our landscape embodies all the uncertainty of a storm, suddenly hitting a mountain, a coastal city, or a tree, shaking the very rootedness of our survival on Earth.

Large infrastructural projects like dams and dikes ignore—or rather counter—the inherent dynamics of natural systems with the aim of rendering them more static and predictable. While this approach has created a seemingly safe environment, it has implied a decrease in ecological value and an increased risk of natural hazards. New approaches in landscape design that tend to analyze such large-scale morphological processes include landscape urbanism,[1] landscape infrastructure,[2] and ecological urbanism.[3] In this case, ecological systems are understood as a transformation of fields and flows and are preferred to the formal design arrangement of fixed objects. It has become clear over the past two decades that one cannot design or control the landscape environment completely; a more symbiotic, all-encompassing approach to landscape processes is necessary to attain more lasting and resilient landscape developments. As such, the belief in a "makeable" designed landscape is shifting drastically.[4] New approaches to landscape design highlight decentralized, small- to medium-sized projects that embrace local materials and processes

---

1. Charles Waldheim, "Landscape as Urbanism," in *The Landscape Urbanism Reader*, New York: Princeton Architectural Press, 2006.
2. Pierre Bélanger, *Landscape as Infrastructure: A Base Primer*, Abingdon, New York: Routledge, 2017.
3. Mohsen Mostafavi, Gareth Doherty, *Ecological Urbanism*, Zurich: Lars Müller Publishers, 2016.
4. The term "makeable" is a loose translation from the Dutch term *maakbare* or *maakbaarheid* that indicates an analysis of the possibility for constructing an object. It is often used in the Netherlands relating to landscape and the extent to which we can make (and control) it. The belief in the *maakbare* landscape found its peak in the post-World War Two era, see Marinke Steenhuis, Fransje Hooimeijer, *Maakbaar Landschap: Nederlandse Landschapsarchitectuur*, 1945–1970, Rotterdam: NAi Publishers, 2009.

and point towards a new understanding of how we operate responsibly within our environment.

THE ARCHITECT As Antoine Picon notes, "It is strange to observe how the discourse on emergence has so far been monopolized by architects and architectural theorists although it applies in a more obvious way to landscape than to architecture."[5] The current landscape condition can partly be ascribed to the development of powerful hydraulic equipment in the 19th century, which changed the way local materials were sourced, leading to traditional construction techniques being applied at the scale of an entire territory. The ease of mechanical material manipulation and transportation gradually replaced the value of handicraft and careful manual assembly that had previously prevailed. Before the advent of this mechanical power, minimizing earth movements was more of a necessity than an ambition. It required a thorough knowledge of the lay of the land, its topography, water regimes, and soil build-up. Without economical constraints in earthmoving, a site could be entirely modified and transformed without much regard for local materials or natural processes.[6] As current large-scale landscape construction practices are now primarily driven by hydraulic equipment, they are often accompanied by extensive material displacement to and from a site.[7] Combined with the global proliferation of powerful grading equipment, this has led to a uniform approach to the shaping of the land, with standardized infrastructural practices

---

5. Antoine Picon, "Substance and Structure II: The Digital Culture of Landscape Architecture," in *Harvard Design Magazine*, 36, 2013.
6. Use of the word "economical" is deliberate, pointing towards a careful handling of material resources.
7. Jane Hutton, "Material as Method," in *Landscript 5: Material Culture: Assembling and Disassembling Landscapes*, Berlin: Jovis, 2017.

transforming the landscape into earthworks.[8] Engineering practice has become so normative and systematic that it has neglected the intrinsic value of each site, leading to a loss of place in terms of performance and meaning.

Beyond the blind faith in the scientific domination of natural processes, we know that making lasting landscapes that are robust and resilient is a difficult quest. Our objective must shift towards considering landscape design as a set of dynamic natural systems that work together. Natural processes that impact a terrain can be traced back to erosion, transportation, and accumulation of granular material, rocks, and organic sediment. Larger natural landforms, while seemingly static, actually undergo continual transformation. Landscape forms should, therefore, be understood as structurally stable moments within an evolving field.[9] The behavior of these natural processes does not lead to a predetermined outcome; instead, their properties are best described statistically through variables and probabilities. This becomes most apparent when we look at the many different time scales that a site embodies, ranging from geological time as in the formation of mountains, up to the rapid changes that often occur through cataclysms and hazards.

Adapting landscapes to continual change seems to be a more fitting approach for an architect in today's climatic context. The idea of terrain representing a stable and permanent entity is gradually shifting away towards a mindset of learning how to deal with constant change. Landscapes, there-

8. In comparison to architectural discourse, the International Style was criticized by Kenneth Frampton towards the end of the 20th century for reintroducing local traditions and aspects of "ground" within architectural projects, against the globalization of architectural form. See Kenneth Frampton, "Towards a Critical Regionalism: Six Points for an Architecture of Resistance," in *The Anti-Aesthetic: Essays on Postmodern Culture*, Port Townsend: Bay Press, 1983, pp.16-31.
9. Sanford Kwinter, "Landscapes of Change: Boccioni's 'Stati d'animo' as a General Theory of Models," in *Assemblage*, no. 19, 1992, pp.50-65.

fore, should be understood by the Architect as open systems influenced by strong dynamic fields and forces in the environment, making their evolution all the more unpredictable. As these natural processes operate in a nonlinear fashion, a convergence of formal and ecological attitudes in design has become a necessity.[10]

With the advent of autonomous robotic earthmoving equipment that operates explicitly through successive procedures over time, it is clear that form only represents the ecological systems of a given place at a particular moment in time.[11] Therefore, design forms and processes are inherently combined in the progressive evolution of a terrain. While the underlying form of a landscape directs the flow of water and earth, so natural processes in turn change the actual form of a place by shifting towards an equilibrium that is constantly being reassessed and redefined. By focusing on the performance of a given landscape, both as an ecological system and as a place for people, the Architect works towards new possibilities that will arise between the forces of society and the forces of nature.

THE ROBOT The Robot's ancestor, the automaton, takes us back beyond the Mechanical Revolution to the farthest reaches of ancient civilizations around the world.[12] For example, it is said that early automatons in the shape of humans were developed in China under King Mu of Zhou, 1,000 years before the Common Era, and there followed more in India and Arabia

---

10. Rod Barnett, *Emergence in Landscape Architecture*, London, New York: Routledge, 2013.
11. See note 9.
12. The oldest known automaton in the Western world is said to be that of the flying dove made of wood by Archytas of Tarento in 400 BCE. See Adrienne Mayor, *Gods and Robots: Myths, Machines, and Ancient Dreams of Technology*, Princeton: Princeton University Press, 2018.

thereafter.[13] Later in the Mannerist Renaissance, automatons were created for the amusement and wonder of visitors to private gardens where rudimentary robots, activated by water, performed enchanting melodies and ritual dances. These robots were not meant to perform menial garden tasks but rather to awaken wonder and admiration.[14] Later, through the development of elaborate clockwork mechanisms, the automaton came to embody the modern spirit of precision and time.

The word "robot" appeared as recently as the 20th century, stemming from the Czech word *robota* meaning forced labor.[15] "Robot" is, therefore, a term that shows the darker side of things—of a world forced into mechanical existence and directed by machines. The OED defines a robot as "An intelligent artificial being typically made of metal and resembling in some way a human or another animal."[16] The notion of the robot as an intelligent, self-regulated device appeared from the Industrial Revolution onward in a multitude of forms as large as human imagination. Devil or angel, more prosaically, the robot has now become ubiquitous in most factories and assembly plants around the world, taking over the most tedious tasks with high precision. What is new, however, is the robot's newly acquired independence from human control and intelligence. We have reached a point where the self-driven car, truck, or bulldozer will become part of daily reality.

13. Legend has it that Yan Shi made a moving figure of wood and leather resembling a man, astonishing King Mu of Zhou. See Joseph Needham, *Science and Civilisation in China, Volume 3, Mathematics and the Sciences of the Heavens and the Earth*, Cambridge: Cambridge University Press, 2005.
14. In Salomon De Caus's *Les Raisons des Forces Mouvantes avec Diverses Machines*, published in 1615, one finds the illustration of a musical fountain comprised of a moving dove and owl, for instance.
15. The word "robot" first appeared in Karel Čapek's 1920 play, *R.U.R.* (*Rossum's Universal Robots*).
16. "Robot, n. 2," in Oxford English Dictionary online, https://www.oed.com/view/Entry/166641?rskey=BGVtdS&result=2 (accessed Feb. 2, 2021).

In robotic landscape fabrication, a new terrain structure is achieved through combined computation and mechanization. The self-organization of granular material helps form a new topology and landform resulting from the dynamic use of robotic actuation. While the tectonic in architecture can be described as the art of expressing forces acting on a material system, topology in landscape architecture may now be characterized as the art of expressing these processes at the surface of the terrain.

In designing a set of dynamic procedures over time, a new landscape form emerges from a dialogue between the designer, natural processes, and the robotic manipulation of terrain. Ironically, the most uncertain condition lies in the evolving insights and demands of economic or social origins. While robotic processes can provide continuous landscape maintenance in response to on-site events, a design goal initially defined by the designer at the moment of conception will inevitably change due to progressive insights and societal priorities. As such, robotic construction processes force a shift—from designing that which is already known to that which is still unknown—thus allowing for countless future readjustments. This shifts the object of landscape design towards a time-based conception: from *what* needs to be achieved to *when* it needs to be achieved. As a landscape will never stop eroding and changing, in that respect landscape design should equally be continuously questioned and transformed.

The question that remains for the Architect is how to harness this modern robotic beast back into the service of society and intelligent design. Through the use of computation and on-site robotic construction processes, landscapes can be informed not only by natural forces but by societal needs and mechanical actuation as well. While research into using robotic systems for governing and altering landscapes exists, those systems have never been mobile.[17] Computational methods inform a designer about the evolution of local natural systems, while mobile robotic platforms enable them to engage in a dynamic response to ongoing processes in the terrain. This opens up an entirely new way of designing, especially with regards to the control of evolving landscape processes.

On the Gürbe River in the canton of Bern in Switzerland, the experiments undertaken by the Robotic Landscapes design research studio at ETH Zurich show how static infrastructure like concrete river channels, steps, and dams can be gradually replaced by a soft, open topography made of local sand, gravel, and boulders. Here, robotic forces modify and transform terrain in response to ongoing site dynamics. This is made possible by the continuous monitoring and analysis of landscape processes where robotic earthwork and maintenance strategies provide the necessary levels of control, always keeping the landscape in dynamic equilibrium. The changing terrain structures operated by the robot not only counteract natural forces and provide safety, they also leverage

17. These robotic systems are mostly oriented towards the direction of water bodies, see Bradley Cantrell, Zihao Zhang, "A Third Intelligence," in *Landscape Architecture Frontiers* 6, no. 2, 2018, p.42.

erosion, transportation, and sedimentation processes to their advantage, strategically aligning with—or opposing—flows of material. As such, this approach is more aligned with designs that offer adaptive approaches over finalized geometric schemes. While the necessity for maintenance only comes from social interest in the protection of a certain territory, landscape interventions can equally be tuned to respond to new needs over time. Beyond the specific design problem linked to a given site, this iterative, adaptive, computationally informed process helps develop new, stronger competences in the Architect. The flexibility and capacity to integrate change and complexity through robotics with subtlety, finesse, and intelligence is exactly the skill set new architects need.

CONCLUSION The Storm, the Architect, and the Robot propose a geometric equation for a new kind of design approach in the making. One where the Storm, representing the forces of nature computationally modeled, is repeatedly integrated, tested, and simulated throughout successive design iterations. The Robot, on the other hand, proceeds parametrically through a project data set; with the help of its sensors and drones, it computationally interprets the reality of a terrain to be worked on and modified. Through a repeated series of computational procedures, a robotic device—like the self-operating excavator HEAP—combines real-time information about the terrain and conflates it with the operating procedures that have been fed in by the designer.[18] The

18. For a description of HEAP, see Dominic Jud, Marco Hutter, "Autonomous Earthmoving," later in this book.

Architect defines design targets and options from an array of societal options and constraints, interprets the robotic data and results, and further guides and refines the output each step of the way. It is this new capacity to negotiate the form of a design among shifting goals, stakeholders, and forms that will help define an entirely new form of landscape aesthetic to serve our society, one that will be closer and truer to the reality of a site by delivering designs capable of greater resilience tomorrow.

We urgently need to develop a more sustainable and resilient form of natural environment, one where our design approach must align itself with dynamic processes that have shaped the place. Through the combination of robotic technologies and computational tools and design, a new landscape reality is created where a continual adaptation of terrain becomes a reality. In this instance, landscapes are formed by both cultural, natural, and mechanical forces in procedures that evolve over ever-changing social and environmental conditions. Instead of making the form of a landscape in design, we should rather learn how to maintain it with the combined help and talent of the Architect and the Robot. This might just be the right time to reconsider the motto on the Netherlands' coat of arms and apply it to our practice on terrain: *Je Maintiendrai*—"I will maintain."

FOUNDATION

A shift from conventional to robotic modes of construction will require a dynamic new design methodology. *Foundation* outlines these new methods wherein a close connection between physical and digital models establishes a more informed design and construction practice. Five texts by Dominic Jud, Marco Hutter, Mathias Bernhard, Ilmar Hurkxkens, Benedikt Kowalewski, Ammar Mirjan, Jesús Medina, and Fujan Fahmi give an overview of the conceptual and practical prerequisites in design. These help translate robotic techniques applied in early robotic sandbox experiments into effective implementation strategies in a risk-prone Swiss Alpine context. Design actions here are inherently linked to topological intelligence, where landscape performance can be thought of in terms of surface and understood as the result of unique local conditions. Hence, it reinforces the link between evolving terrains, dynamic earthmoving operations, and subsequent topographic transformations.

The full-scale robotic HEAP excavator can test and prove the feasibility of autonomous on-site construction methods through the creation of highly complex terrain geometries using granular materials. This is at the heart of design research experimentation with autonomous robots, forming the starting point for a fundamental shift in our approach to terrain and the resulting construction techniques. With this robotic machine in mind, what are the methods we can employ to reinforce the specificity of a site in a studio setting?

Progress in digital surveying techniques has brought all topographic intricacies pertaining to a landscape back into the studio where they can be assessed visually, topologically, and computationally. To encode design methods for this robotic platform, digital terrain modeling tools were developed that facilitate a more dynamic approach to construction. Besides enabling continual adaptations to the digital model in place, these tools also incorporate simulation software that can test the environmental performance of a wide range of large-scale landscape design proposals. This in turn shifts the emphasis on repeated simulation possibilities in the process of design.

The study of dynamic terrain formations—between material behavior and robotic response—can be further explored using model-scaled sandboxes. This digital-physical "sketchbox" allows experiments in a studio to be set up by integrating data from various sensors on the robotic arm. Thus, dynamic earthmoving tasks can be explored by iteratively scanning and manipulating the sand repeatedly. Through this topological process, highly articulated geometries emerge from these design experiments and simulations, concretely exemplifying how new landscape topographies can be imagined in real-world settings.

The continuous design dialogue between existing and proposed sites—where data at the scale of the site flows into the studio sandbox and back out again—proves to be the fundamental core of a new dynamic design method. It calls on us to respond appropriately to the ever-changing site conditions and new environmental realities that we will face. Embracing dynamic robotic design challenges shows how we can move beyond the status quo in landscape design and engineering, away from preconceived static responses, and towards more adaptive topographic solutions.

AUTONOMOUS EARTHMOVING

DOMINIC JUD
MARCO HUTTER

FIG. 1

HEAP—THE AUTONOMOUS WALKING EXCAVATOR As a proof of feasibility, and to push the state of the art in construction machine automation, the Robotics Systems Lab at ETH Zurich started a research project with multiple industry partners lasting over half a decade to automate the most versatile existing mobile machine on construction sites. We chose a walking excavator over tracked or wheeled excavators due to its superior mobility.[1] The Hydraulic Excavator for an Autonomous Purpose (or HEAP) is the world's first autonomous walking excavator with a high potential for automation in construction with various applications. HEAP is based on a twelve-ton, commercially available Menzi Muck M545 walking excavator, but highly customized with numerous adaptations and additions.[2] A broad suite of sensors was installed, including light-detection and ranging sensors (lidar) to perceive the environment, as well as GNSS-RTK antennas and inertial measurement units (IMUs) to localize the machine on the construction site. As the off-the-shelf machine is built for an operator in the cabin, the actuation principles also had to be changed to focus on high-performance control through an installed computer. Electric pilot-stage valves and high-performance servo valves were added for the arm, while all the cylinders in the chassis were exchanged for servo valve-operated cylinders that allow for force control. These changes have led to a heavy-duty mobile manipulator for various construction tasks.

ROBOTIC EXCAVATION SYSTEM The robotic process for fabricating free-form embankments is split into

FIG. 1 EXECUTION OF THE DUMPING STEP AS PART OF THE ROBOTIC FABRICATION PROCESS OF FREE-FORM EMBANKMENTS WITH HEAP.

1. Dominic Jud, High-Accuracy Autonomous Excavation of Free-Form Shapes (PhD Thesis), Zurich: ETH Zurich, 2021.
2. Dominic Jud, Martin Wermelinger, Simon Kerscher, Edo Jelavic, Pascal Egli, Philipp Leemann, Gabriel Hottiger, Marco Hutter, "HEAP – The Autonomous Walking Excavator," submitted to *Automation in Construction*, 2020.

three major parts.[3] First, a drone performs an initial survey to gather a three-dimensional map of the construction site before design algorithms compute the desired shape of the embankment based on this recorded data. Lastly, the desired elevation is sent to HEAP, which runs at a feedback rate of 100 Hz, to execute the necessary excavation and dump actions to turn the design into reality. The elevation map is the feedback from the fabrication to the design, the rate of which can vary according to the project, as faster rates might be helpful to adapt the design quickly and regularly to changes in the environment. At the other extreme, there can also be no feedback taken from the fabrication to the design, in which case HEAP will try to recreate the initially planned design as well as possible. The drone survey should be used at the beginning to gather a complete map of the environment for the design step, and it could also be discretely deployed to get a complete overview of progress and the possibility of adapting the design, as the excavator's viewpoint might make it miss certain places due to obstructions.

A single cycle of the fabrication process consists of an excavation step and a dumping step. The image shows a half-finished embankment with the finished terrain at the front and an unfinished part at the back. An excavation planner finds the next point of attack to excavate by considering the stability of the embankment as well as the feasibility of a certain location to be excavated. A trajectory free of terrain collisions is planned for the excavator to fill its bucket, after which

FIG. 2

FIG. 2 FIELD EXPERIMENTS WITH HEAP, THE AUTONOMOUS WALKING EXCAVATOR BASED ON A STANDARD MENZI MUCK M545.
FIG. 3 EXTENSIVE HARDWARE CHANGES ARE INTEGRATED ONTO HEAP TO ACHIEVE THE DIGGING AND DUMPING CYCLES. THIS IMAGE ILLUSTRATES:
(1) WIFI ANTENNA
(2) GNSS-RTK AND IMU
(3) LIDAR
(4) TRAJECTORY FREE OF TERRAIN COLLISIONS
(5) PLANNED POINT OF ATTACK
(6) PLANNED DUMP POINT
(7) DRIVING PATH
(8) ACTIVE CHASSIS-BALANCING SYSTEM
(9) FINISHED PART OF THE EMBANKMENT WITH HEIGHT LINES FOR BETTER VISUALIZATION
(10) UNFINISHED TERRAIN

3. Dominic Jud, Ilmar Hurkxkens, Christophe Girot, Marco Hutter, "Robotic embankment: Free-form autonomous formation in terrain with HEAP," in *Construction Robotics*, Springer, 2021.

the dump point planner finds a suitable dump location to further build up the embankment in a safe and stable manner. As the embankment is much larger than the reach of the excavator, the machine is driven along a predefined path to reposition itself. During driving, the active chassis-balancing keeps the excavator's base horizontal and all four wheels on the ground. The final state of the s-curved embankment shows an average error of only 0.056 m, which is remarkably low for such a free-form shape. In general, embankments could be built as wide as five meters and to a maximum height of six meters.

FUTURE DEVELOPMENTS In our demonstration of autonomously built free-form embankments, only the terrain shape was fed back from the fabrication process into the design environment. It allowed for the inference of geometric properties, for example cut and fill volume balancing. Future robotic excavators will allow for a much broader range of feedback that is not only of a geometric nature, but also includes properties such as soil composition, humidity, and obstacle detection (large rocks and roots, for example) that could be fed back into the design. This development will enable the use of local materials for construction, eliminating the need to bring specific additional material to the site or remove truckloads of excess material.

FIG. 3

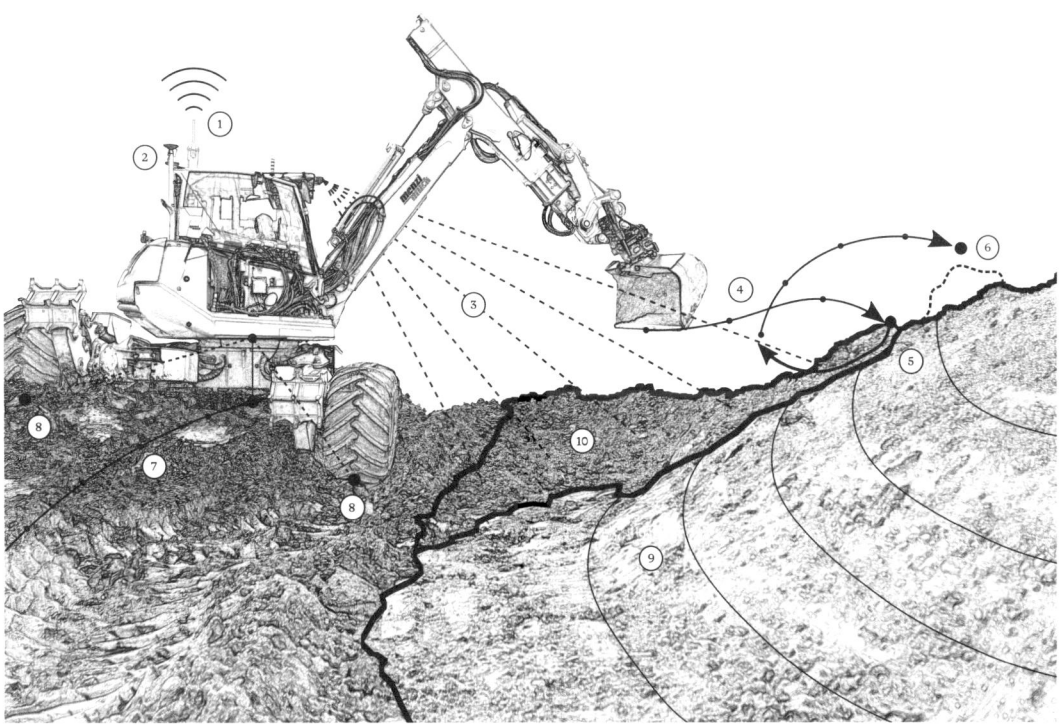

F I G . 4

F I G . 4 THE ONLINE INTEGRATION OF TOPOGRAPHIC DESIGN AND CONSTRUCTION WAS DEMONSTRATED IN A FULL-SCALE EMBANKMENT PROTOTYPE. THE TOP VIEW SHOWN HERE HIGHLIGHTS THE CURVED DESIGN AND SMOOTH SURFACE FINISH.
F I G . 5 VISUALIZATIONS OF A ROBOTIC EXCAVATOR BASED ON HEAP, WHERE HUMAN-OPERATED CONTROL SYSTEMS ARE NO LONGER NECESSARY ONBOARD. HEAP IS 5.83 METERS LONG, 2.37 METERS WIDE, 2.55 METERS HIGH, WITH FOUR CYLINDERS, A 4,100 CUBIC CENTIMETER DISPLACEMENT TOOL, AND A 115 KILOWATT / 157 HORSEPOWER ENGINE.

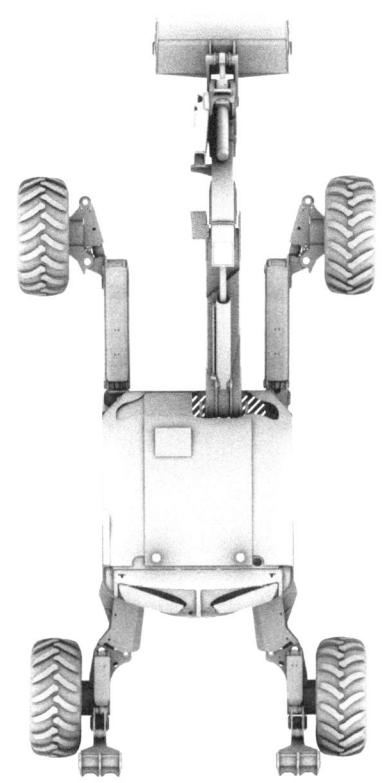

FIG. 5

MATHIAS BERNHARD
ILMAR HURKXKENS

FIG.6

MODELING THE FIELD

1. Benjamin Lee Whorf, "Science and linguistics," in *Language, Thought and Reality: Selected Writings of Benjamin Lee Whorf*, Cambridge: MIT Press, 1940, pp.207–219.

FIG. 6 TWO ILLUSTRATIONS OF A TWO-DIMENSIONAL DISTANCE FIELD WITH THE SAME GEOMETRY DEFINED AS A SIGNED DISTANCE FUNCTION (SDF) WHERE SPACE IS REGARDED AS A FULL, CONTINUOUS FIELD OF VALUES.

ENCODING Computational design tools have become an indispensable part of design processes at every scale, from medical implants to landscape architecture and urban planning. They go beyond being a mere digital drafting board to fulfill tasks like animation, simulation, fabrication, and even the generation of new designs. For a computer to understand a concept and allow for its modification, it needs to be made machine-readable, for which we use the term "encoding." This translation to a digital representation determines the degrees of freedom a designer has, the number and types of possible operations, and the size and shape of the design space. The tools design professionals work with have an undeniable influence on the result of the creative process. Computational design can be an instrument rather than a tool as it depends on the virtuosity of the one playing it, instead of just being a problem solver. It changes how we model, and it has the potential to fundamentally change what we model. As Benjamin Lee Whorf puts it, "Language shapes the way we think, and determines what we can think about."[1]

TERRAIN Architectural construction can be thought of as assembling elements (bricks, prefabricated walls, or other parts) in midair, in empty space. The virtual model for planning most often represents these elements as geometric primitives consisting of points, curves, and surfaces. For landscape architecture, the context and conditions are different; the terrain is already present, has an organic form, and the space is either a solid substance

2. A geographic information system, or GIS, is a digital framework providing tools to read, store, edit, and analyze spatial and geographic data.
3. James F. Blinn, "A Generalization of Algebraic Surface Drawing," in *ACM Transactions on Graphics*, 1 (3), 1982, pp.235-256.
4. Jules Bloomenthal, Bajaj Chandrajit, "Introduction to Implicit Surfaces," in *The Morgan Kaufmann Series in Computer Graphics and Geometric Modeling*, San Francisco: Morgan Kaufmann Publishers, Inc., 1997.
5. Alexander Pasko, Valery Adzhiev, Alexei Sourin, Vladimir Savchenko, "Function Representation in Geometric Modeling: Concepts, Implementation and Applications," in *The Visual Computer*, 11 (8), 1995, pp.429-446.
6. Mathias Bernhard, Michael Hansmeyer, Benjamin Dillenburger, "Volumetric Modelling for 3D Printed Architecture," in *AAG 2018, Advances in Architectural Geometry*, Vienna: Klein Publishing GmbH, 2018, pp.392-415.
7. William E. Lorensen, Harvey E. Cline, "Marching Cubes: A High Resolution 3D Surface Construction Algorithm," in *Proceedings of the 14th Annual Conference on Computer Graphics and Interactive Techniques - SIGGRAPH 1987*, 21 (4), 1987, pp.163-169.
8. Wiki McNeel, "Why Booleans Fail," https://wiki.mcneel.com/rhino/booleanfaq (accessed July 16, 2019).
9. Ilmar Hurkxkens, Mathias Bernhard, "Computational Terrain Modeling with Distance Functions for Large Scale Landscape Design," in *Journal of Digital Landscape Architecture*, Berlin: Wichmann Verlag, 2019.

or a void. Operations are more likely to modify and deform existing material rather than assemble new discrete elements, hence their encoding must be reconsidered to find the best and most versatile computational representation for working with different terrains. Conventional CAD tools originating in the transportation industry (for ships, cars, and airplanes) are unlikely to be the most well-suited.

A terrain has an amorphous form and can therefore only be poorly approximated with geometric primitives, similar to how a sunset can only be poorly drawn with a vector graphic. Encoding similar to that used for photographs (raster graphics) holds the promise of flexibility, while orthogonal grids of elevation values are already a very common format for GIS applications.[2] We choose to encode terrain as a discrete field of values, modification operations as continuous mathematical functions, and shapes and forms as calculations instead of constructions. Functions are completely free of resolution and can be rendered at different discretization intervals.

INSTRUMENT A simple rectangle can be encoded either as a list of corner vertex coordinates and their connections with lines, as the iso-value contours of a function, as the sum of odd fractions of sine waves, or in arbitrarily many other ways. Many different representations compete for being the best one for specific purposes, but we found one not based on points, curves, and surfaces but on functions to be more versatile and suitable for representing landscape forms. The principle has been known for a long time but has

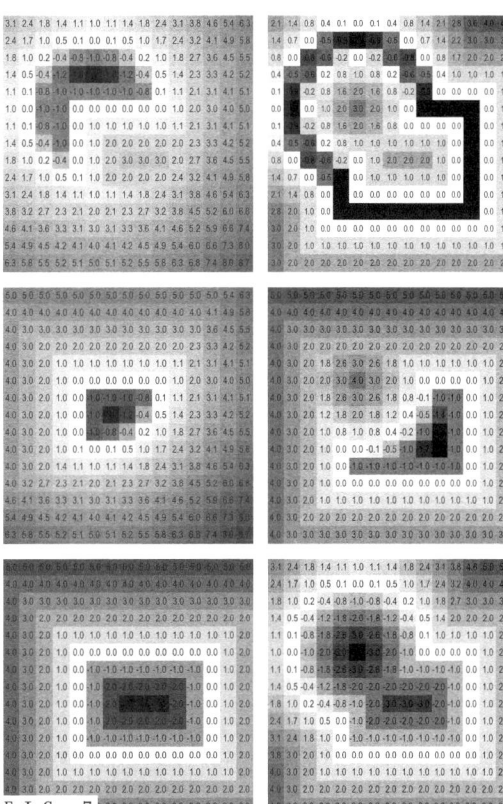

FIG. 7 A GRAPHIC AND NUMERIC DEMONSTRATION OF TWO PRIMITIVES—A RECTANGLE AND A CIRCLE—AND VARIOUS BOOLEAN OPERATIONS BETWEEN THE TWO GEOMETRIES.

a less prominent role in the history of computer graphics;[3] its nomenclature knows many alternatives, implicit as opposed to parametric modeling,[4] or function representation (FRep) as opposed to boundary representation (BRep).[5] We call it volumetric modeling (VM)[6] as it also comprises methods for voxel manipulation learned from medical imaging.[7]

In conventional CAD software, points are placed in empty space, connected by curves and eventually surfaces. The designer constantly has to take care of the correct geometric topology—the connectivity of these elements—but with increasing complexity, certain operations are prone to fail.[8] With VM, we describe forms mathematically as signed distance functions (SDFs). These functions can have geometric features as input, but can also be abstract trigonometric or noise functions, and the modifications are purely numeric instead of geometric. For relative operations, a cut is the subtraction of the function result from the field, a fill is its addition. Absolute operations compare the two values and return the lower for a cut and the higher for a fill, allowing us to compute shapes instead of constructing them. The actual geometry comes at the very last step and is basically just one possible type of rendering of the field.

We implement the functions as Python UserObjects for the Grasshopper plugin in Rhino 3D where the basic data structure is a list of numbers consisting of a header (specifying the number of columns and rows, grid spacing, origin location, etc.) followed by the z-elevation values for all the grid points.[9] This

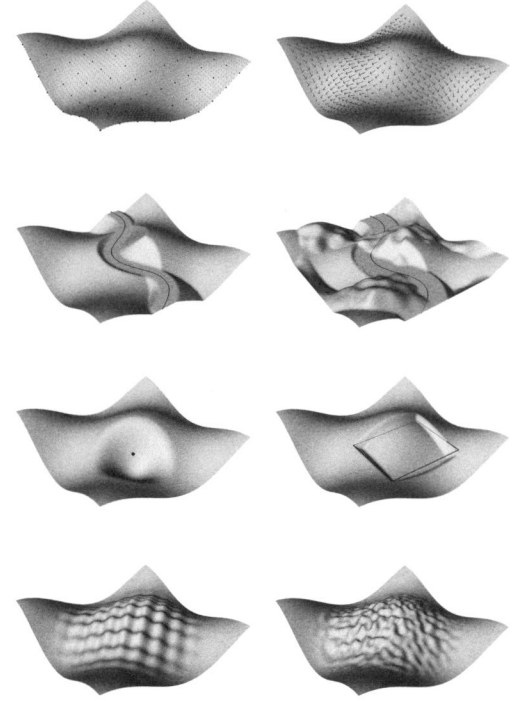

FIG. 8 ILLUSTRATIONS DEPICTING A SELECTION OF TERRAIN MODELING TECHNIQUES APPLIED IN THE ROBOTIC LANDSCAPES DESIGN RESEARCH STUDIO. THESE ARE MODES OF TOPOGRAPHIC TRANSFORMATIONS RANGING FROM GRID TRANSLATIONS AND LANDFORM CLASSIFICATIONS TO CUT AND FILL OPERATIONS ON POINTS, PATHS, AND AREAS.
FIG. 9 ILLUSTRATION SHOWING THE ADVANTAGE OF MODELING WITH NUMEROUS DISTANCE FUNCTIONS IN A DIGITAL TERRAIN MODEL. THE SURFACE IS ADJUSTED USING CONTINUOUS MATHEMATICAL REPRESENTATIONS PERMITTING EXISTING AND DESIGNED LANDFORM FEATURES TO EFFORTLESSLY BLEND INTO ONE OTHER.

FIG. 8

list object is passed from component to component, retrieving the corresponding modifications of the field values based on the results of the SDF. It can eventually be turned into a quad mesh, laid on top of the field like a cloth, along the 0-level, separating solid from void.

MACHINE AND MATERIAL Even though the mathematical principles at work behind the scenes are generic, the terminology "cut" and "fill" are a reference to the main application of the instrument. A cut removes existing material and a fill adds material to the existing terrain. Often it is important to know the difference—the amount of material removed or added by achieving a specific shape—and a modeler based on BReps would have to create a watertight closed shape of the intersection of the states before and after and then calculate the volume. With a method based on a grid of z-values, we can return the arithmetic difference in z for all the affected grid cells, multiplied by their area.

By encoding the modification operations as distance functions, it is equally straightforward to define a region of influence together with a function for how to transition from fully affected to no effect. On small scales, these slopes can represent the shape of a shovel, while on larger scales they constitute landforms and can be strictly linear, smoothly sinusoidal, or any other function to map x in the domain of the region to a corresponding y—the intensity of influence. The possibility to control the angle of these slopes allows work on different terrain materials and different structural properties of a particular site.

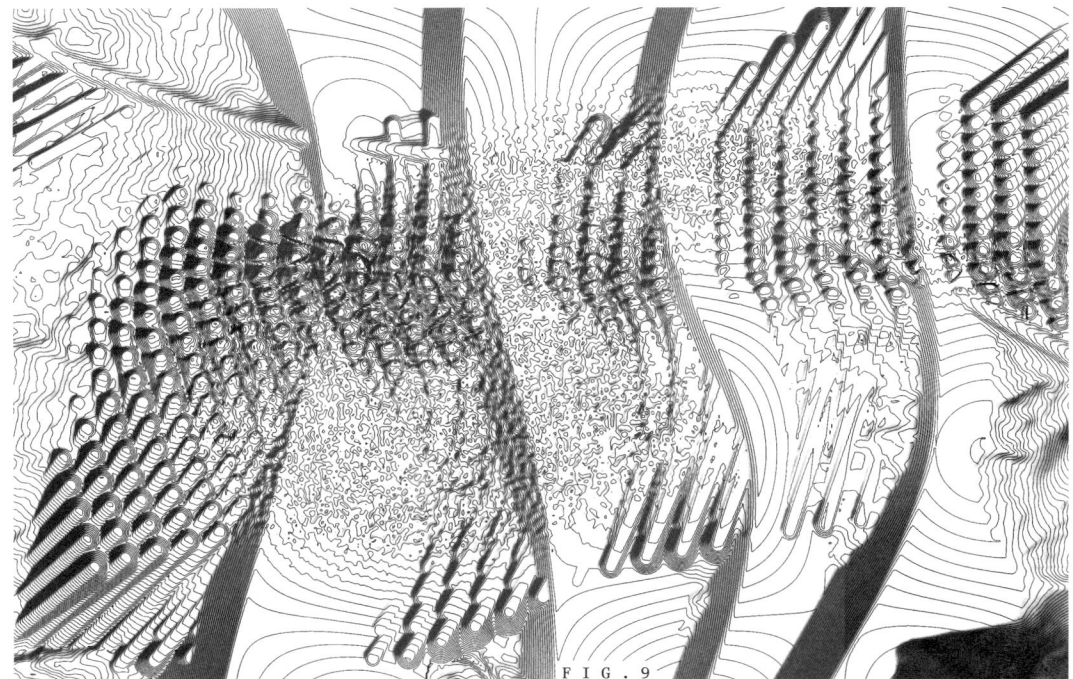

FIG. 9

Transforming elevation values with volumetric methods facilitates interoperability between the topographic survey and the modeling—and fabrication—environment, while providing the possibility to dynamically integrate ever-changing site conditions. As the found terrain already displays an unimaginable richness of geometric complexity, maybe volumetric modeling can provide a good way to encode the shape of landscapes to come.

SIMULATING MATERIAL FORMATION

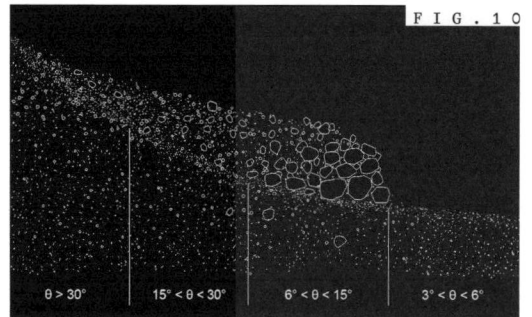

FIG.10 ILMAR HURKXKENS
BENEDIKT KOWALEWSKI

Topographic or bathymetric features visible on or near the Earth's surface are created by the interaction between internal tectonic and external surface processes that are of a physical, chemical, or biological nature.[1] These can be divided into hypogene action (volcanic eruptions, earthquakes) and epigene action (wind, water, life), though all can, in general, be traced back to erosion, denudation, and accumulation. Instead of progressing towards a final equilibrium with occasional local disruptions, it is now understood that natural processes are inherently dynamic, and small disturbances can lead to large effects.[2] The recent increase in natural hazards like sea-level rise, landslides, floods, and drought show us that there exists a delicate balance in our ecosystems. The resulting equilibrium shift we witness today renders many climate predictions of the last decades already outdated. In short, making landscapes that are forever robust and resilient is virtually impossible in light of their dynamic nature. Therefore, ongoing monitoring, simulation, and adaptive strategies will be mandatory for future landscape design.

The Robotic Landscapes design research studio's experiments on Switzerland's Bondasca and Gürbe rivers intended to integrate continuous erosion and deposition processes into a procedural and iterative solution instead of mitigating the damage of inevitable future hazards. Working with the Swiss Federal Institute for Forest, Snow, and Landscape Research (WSL), an iterative design process was established between topographic modeling using volumetric

FIG. 10 THE CHARACTERISTIC LONGITUDINAL SECTION OF A DEBRIS FLOW. SEDIMENT SURGES DOWNHILL IN RESPONSE TO GRAVITATIONAL ATTRACTION, WITH LARGER PARTICLES (BOULDERS) AT THE FRONT OF THE MASS WHILE SMALLER PARTICLES COME BEHIND, ITS LONG TAIL OFTEN SATURATED WITH WATER. A DEBRIS FLOW CAN GROW IN SIZE WHILE FLOWING OVER TERRAIN DUE TO BED EROSION CAUSED BY LARGE BOULDERS.

1. Philip A. Allen, "Time Scales of Tectonic Landscapes and Their Sediment Routing Systems," in *Geological Society, London, Special Publications*, 296, no. 1, 2008, pp.7–28.
2. Jianguo Wu, Orie L. Loucks, "From Balance of Nature to Hierarchical Patch Dynamics: A Paradigm Shift in Ecology," in *The Quarterly Review of Biology*, 70, no. 4, December 1995, pp.439–466.

methods[3] and rapid mass movement simulation.[4] This new workflow enabled the integration of site-specific topographies instead of relying on general rules relating to erosion and deposition for alpine rivers. As such, it was possible to model the influence of landslides and debris flows on proposed designs over a more extended time period.

In the case of the Bondasca, three phases were chosen that align to expected future events amounting to 400,000, 150,000, and 50,000 m³ respectively, while at the Gürbe river, a yearly deposition of 50,000 m³ from the upper valleys was expected.[5] As debris flows and mountain torrents incessantly transport sediment, new granular material is continuously introduced into the study perimeters of the rivers' respective valleys. The introduction of sediment occurs either in major events such as debris flows, or more gradually over several years by river dynamics. The intertwining of volumetric modeling methods and rapid mass movement simulation was used throughout the experiments to understand these quasi-infinite cycles of natural processes and design interventions. Therefore, this workflow was not only able to validate the final product, but rather informed a continuously evolving design. The experiments show how to adapt the topography and anticipate a continuous evolution over an extended period of time through multiple iterations, thereby assessing its resilience to uncertainties.

To allow for an intuitive design process, a common data structure for both the terrain modeling and

FIG. 11 SLOPE GRADIENT ANALYSIS OF THE EXISTING SITUATION ON THE GÜRBE, WHERE THE RIVER CHANNEL IS RENDERED IN A MULTITUDE OF HORIZONTAL SECTIONS BY THE MANY CONCRETE CHECK DAMS (DARKER COLORS INDICATE FLAT AREAS, WHILE LIGHTER COLORS SIGNIFY STEEP SLOPES).

3. For elaboration on digital terrain modeling methods, see Mathias Bernhard, Ilmar Hurkxkens, "Modeling the Field," earlier in this book.
4. Marc Christen, Yves Bühler, Perry Bartelt, Remco Leine, James Glover, Adrian Schweizer, Christoph Graf, et al, "Integral Hazard Management Using a Unified Software Environment: Numerical Simulation Tool 'RAMMS' for Gravitational Natural Hazards," in Conference Proceedings, Grenoble, France, International Research Society, Klagenfurt, 2012, pp.77–86.
5. The project around the Bondasca river retroactively designed a new detention basin as the existing one had failed dramatically during a mudslide in 2017, explaining the necessity of these large numbers. For the Gürbe river, the project's objective halted all maintenance checks of the dams, resulting in high yearly volumes of material stemming from the riverbed itself. See Christian Wilhelm, Gian Cla Feuerstein, Andreas Huwiler, Roderick Kühne, "Bergsturz Cengalo und Murgänge Bondo: Erfahrungen der kantonalen Fachstelle," in *Lernen aus Extremereignissen*, 2019, p.53.
6. The grid data for the surrounding digital terrain model (DTM) was sourced from Swisstopo, while local terrestrial laser scans were made to increase local resolution of the lidar data.
7. For showing such a devotion to this project, we would like to thank Brian McArdell, the scientific staff member of the WSL for Mountain Hydrology and Mass Movements as well as Torrents and Mass Movements.

morphological simulation tools was used. Here, regular grid data from aerial as well as terrestrial lidar surveys provided the underlying topographic description to test and design evolutionary scenarios.[6] The specialized simulation tool calculates erosion, transportation, and deposition of sediment for debris flows using soil friction parameters. In order to provide informative results, the software requires mechanical inputs as well as basic parameters such as the frictional resistance of the subsurface, information on the composition, viscosity, and volume of the debris flow, and the topographic model. To ensure real-world results, all parameters of the simulation were coordinated in advance with experts from the WSL.[7] After each simulation, the topography of the design environment was updated to reflect the eroded and deposited sediment in the digital terrain model, and the resulting model thus contained information not only on whether the mitigation of the debris flow was successful, but also where and how much material was moved. At this point, a robotic response was proposed and modeled, which in turn was simulated again to close the loop between material formation based on natural and robotic processes. Thus, the simulation of debris flows was not just a one-time opportunity to check the effectiveness of the intervention, rather, it formed the basis for the next design step. As such, a model generated by the mass movement simulation was just another intermediate step in a never-ending interplay of natural dynamic processes and design interventions.

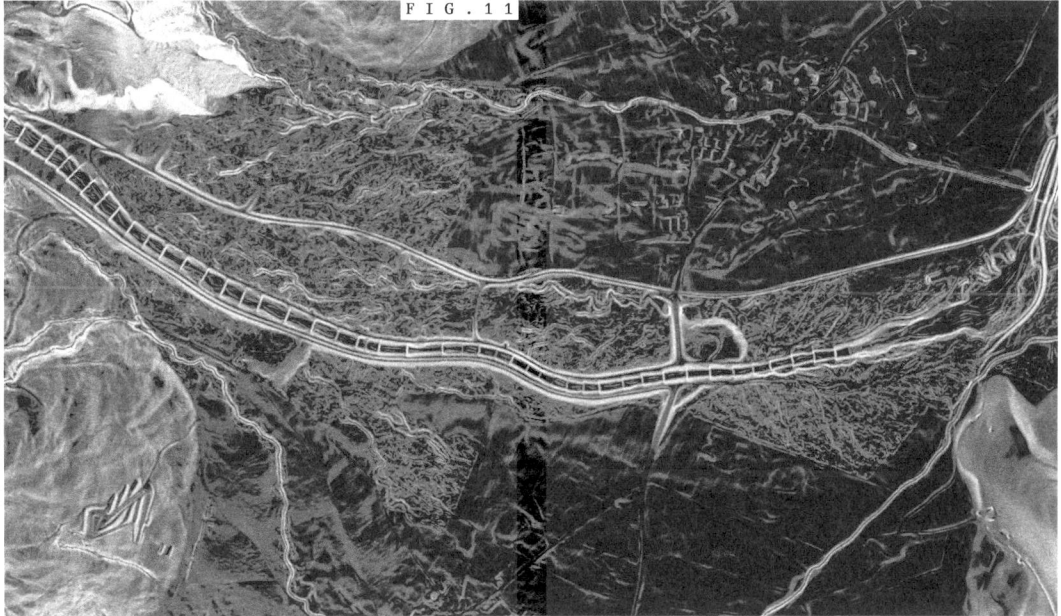

FIG. 11

The modeling approach in the experiments—where a dynamic response to simulated changes occurred automatically in the digital terrain model—is essential for the intended construction method: autonomous earthmoving. A new formation process can be established where simulated deposition combined with earthmoving operations results in creative and unconventional design proposals. However, the true potential of the interaction between simulation and design lies outside of a design studio when large-scale construction machines shape the landscape autonomously. Here, epigene and robotic actions may unfold into a different landscape future than the one pre-envisioned by the designer (whose role will therefore have to be redefined). Instead of designing static moments within a continuous flow of morphological processes, landscape architects will have to become curators of landscape evolution.

FIG. 12 TO STUDY THE CAUSE AND EFFECT OF RAPID MASS MOVEMENT PROCESSES ON TOPOGRAPHY, A RAMMS SIMULATION OF 400,000 CUBIC METERS OF BOULDERS, ROCKS, AND MUD WAS TESTED ON A DIGITAL TERRAIN MODEL, VERIFYING THE SCOPE OF DESTRUCTION OF THE DEBRIS FLOW THAT OCCURRED IN BONDO IN 2017.
FIG. 13 RAMMS SIMULATION OF THE 1990 DEBRIS FLOW EVENT INVOKED BY HEAVY RAINFALL, CAUSING OVER 200,000 CUBIC METERS OF MATERIAL IN THE UPPER REACHES OF THE GÜRBE RIVER TO TRAVEL DOWNSTREAM. THE SIMULATION ENABLED THE ROBOTIC LANDSCAPES DESIGN RESEARCH STUDIO TO DETECT THE AMPLITUDE AND SCALE OF THE NATURAL HAZARD.

FIG.12

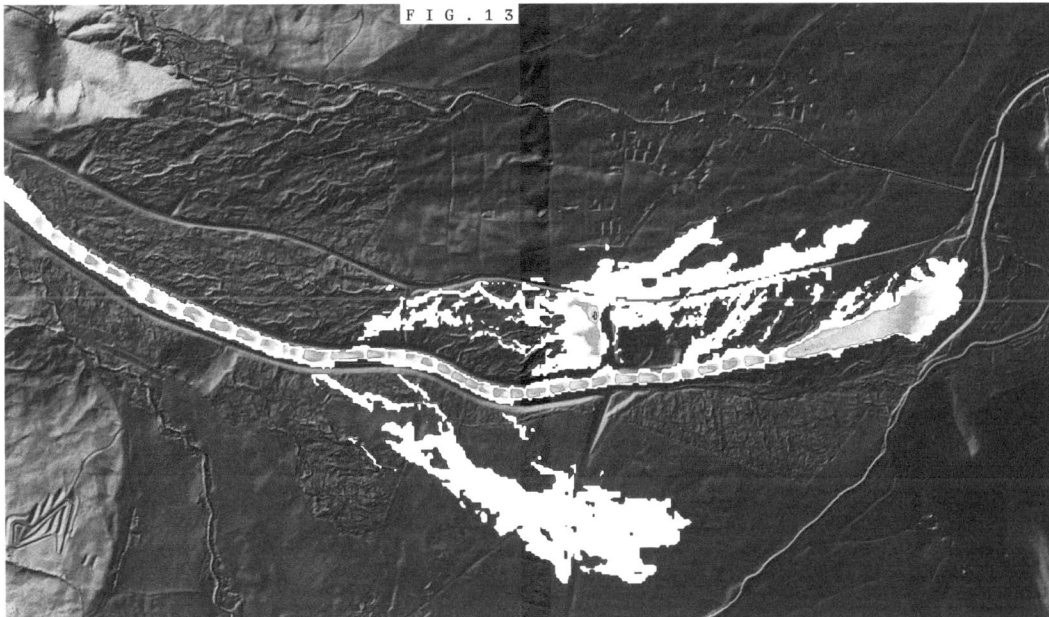

FIG.13

AMMAR MIRJAN
JESÚS MEDINA

FIG.14

THE ROBOTIC SKETCHBOX

1. A set of custom-made Rhinoceros 3D, Grasshopper, and Python components was developed at Gramazio Kohler Research, ETH Zurich, for teaching and research purposes.
2. The UR10 is a collaborative industrial robot arm designed to work alongside humans.
3. For example, it could move along the z-axis until a force of 60 N is reached along the y-axis.

FIG. 14 THE INCORPORATION OF LOCALLY SOURCED MATERIAL IN THE SANDBOX FACILITATED THE STUDY OF UNIQUE LANDSCAPE CONDITIONS, WITH NEW LANDFORMS ARTICULATED THROUGH SITE-SPECIFIC MATERIALITY.

Conceptualizing a landscape intervention by interacting with soil, gravel, or sand presents a special challenge to the designer. The behavior of such granular matter exposed to gravity and friction between the particles often leads to unpredictable material formations. For the designer to approximate and potentially exploit the expressiveness of making landscapes, the Robotic Landscapes design research studio developed a digital design and production process that enables the creation, observation, and ultimately the control over the formation of materials.

The setup consists of CAD components, a robotic arm equipped with different tools, a box filled with granular material, and a 3D scanner. The CAD components—a set of computational design algorithms[1]—enable geometric information from the computer program to be translated into kinematic motion instructions readable by the controller of the robotic arm.[2] The robot executes the motion commands corresponding with the spatial information defined in the CAD; for example, a movement along a spline curve or a 3D surface according to the sequence defined by the designer. Additionally, the wrist of the robotic arm is equipped with a force torque sensor, facilitating the overlaying of spatial information with force feedback, the mechanism of which enables the robot to move along a spatial path until it reaches a specified force in a defined direction.[3]

The wrist of the robot can be equipped with a variety of tools, making it possible to manipulate the material in a range of ways. Depending on the type

4. We implemented a Microsoft Kinect 2 as well as an Intel RealSense depth camera to capture the geometry of the sand.

of material interaction and the design intent—for example shifting, depositing, or compressing—the designer can apply different tools and potentially develop dedicated custom tools. Positioned within the operational reach of the robotic arm, a box may contain different granular materials such as oil-bound sand, gravel, or sand from one of the sites. The 3D scanner can capture and digitalize a material formation within the box and send it to the CAD program as a point cloud.[4] The designer can analyze the three-dimensional data and define algorithmic strategies to further process the material. This fosters the emergence of applicable features while interacting with the materials—for example, specific high and low points or slope angles—as design drivers. The robotic setup enables the direct linking of computational design with material production, and the scanning device feeds back the physical to the digital. The designer is at the center of this loop and interacts with the material formation processes in both directions.

At the core of the robotic design and production process is the box and the material it contains. It is an instrument for prototyping, a mass to shape and reshape. The box is neither a scaled-down segment of the actual site, nor a representation of a potential robotic excavator process. It is an abstraction, an explorative device to steer material formation according to a digital design logic in order to develop a distinct architectural expression. Different from, for example, a milled model, the box does not map a final geometry but a formation at a certain fabrication step

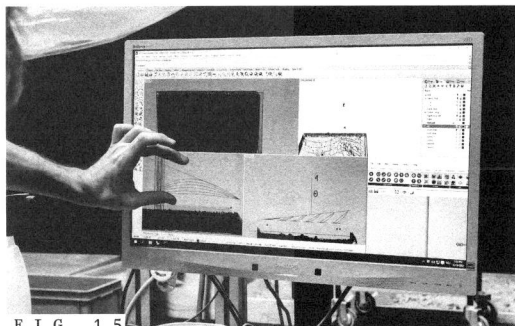

FIG. 15

FIG. 16 BY CREATING AN ALGORITHMIC STRATEGY,
DESIGN INTENTIONS ARE TRANSLATED INTO MACHINE
INSTRUCTIONS FOR THE ROBOTIC ARM.

FIG. 15 USING A 3D SCANNER, THE SANDBOX IS
DIGITALLY CAPTURED AND THEN ANALYZED IN VARIOUS
LOCATIONS RELATING TO THE PROPOSED DESIGN
INTERVENTION.
FIG. 16 BY CREATING AN ALGORITHMIC STRATEGY,
DESIGN INTENTIONS ARE TRANSLATED INTO MACHINE
INSTRUCTIONS FOR THE ROBOTIC ARM.

before it is altered again. Naturally, no material is lost, and the box is therefore a procedural model that is never actually finished. A prototype can be realized in a matter of seconds, encouraging an iterative way of working where the designer can quickly switch between digital design and material execution. As such, the box presents a tool that facilitates a form of sketching with a robot—a spatial drawing paper inscribed by the designer and materialized by soil. A "sketchbox," the observations and learnings from which cultivate reflection on the design of formative material processes based on digital data, fostering a new perspective linking these findings with future construction sites as well as the larger field of landscape architectural design.

FIG. 17 TO CONDUCT DESIGN RESEARCH, A COLLABORATIVE UNIVERSAL ROBOTS UR10 ARM WAS COMBINED WITH ENVIRONMENTAL SENSORS TO MIMIC THE CAPABILITIES OF A ROBOTIC EXCAVATOR. THIS ENABLED THE DEVELOPMENT OF AUTONOMOUS PROCEDURES IN A MODEL-SCALE SANDBOX, RANGING FROM DIGGING AND DUMPING TO SPREADING AND COMPACTION CYCLES.
FIG. 18 VARIOUS END-EFFECTORS WITH DIFFERENT MATERIAL-HANDLING CAPABILITIES WERE DEVELOPED. THE REPEATED IMPRINT OF A SIMPLE TOOL GENERATES COMPLEX GEOMETRIES IN GRANULAR MATERIALS AND INFLUENCES SUBSEQUENT ROBOTIC MOVEMENTS IN A CONTINUOUS PROCESS BETWEEN DIGGING, DUMPING, SHIFTING, AND COMPACTING.

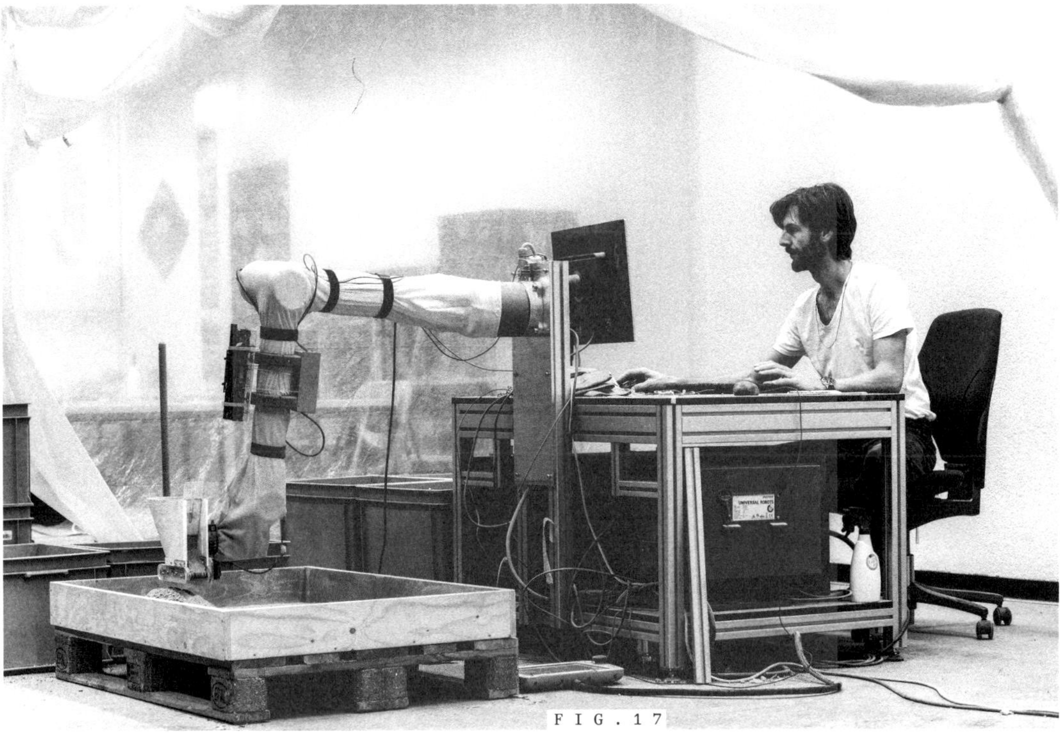

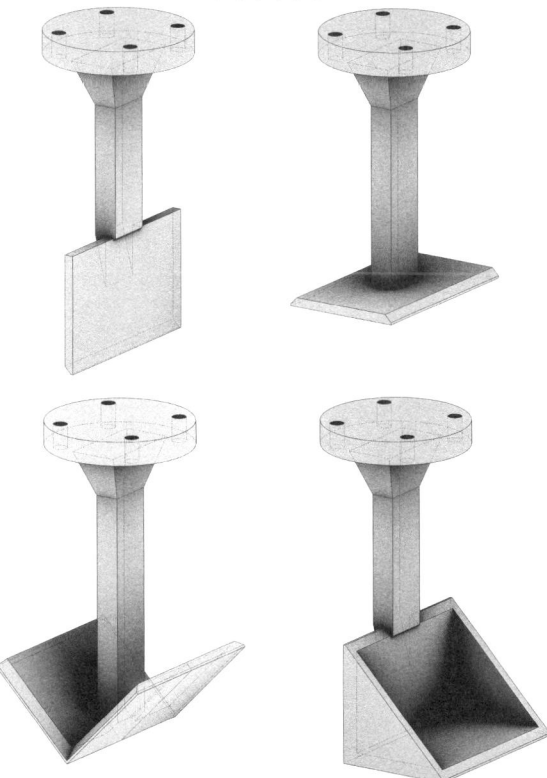

FIG. 17

FIG. 18

TOPOGRAPHIC INTRICACIES
FIG.19 FUJAN FAHMI

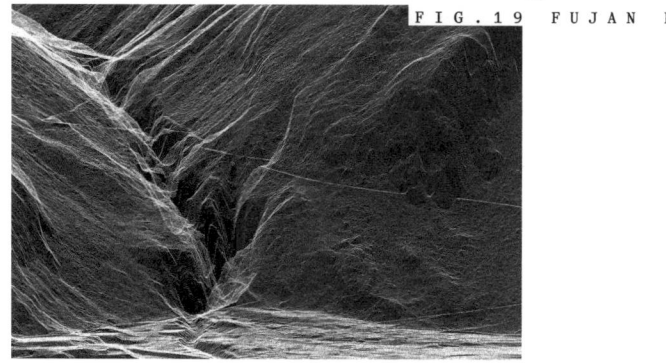

Natural disasters appear in our collective imagination as adverse phenomena. In recent decades, climate change has increasingly gained our awareness as climate records indicate a significant increase in average temperatures and the associated glacial melting in alpine landscapes. This has led to a reconsideration of well-known natural hazards and, above all, to the recognition of new events. Equipped with a precise topographic understanding of a given terrain, natural occurrences can be orchestrated without the need for protective barriers and, in doing so, we can reclaim the unique character of alpine riverbeds as accessible public space.

  The Swiss landscape has been subject to constant change over the past centuries due to major infrastructure projects and flood protection measures. The strategies and configurations required to ensure their effectiveness have also undergone significant changes over the past centuries. Starting in the 18th century, environmental projects such as the Linth river or the first Jura river correction occurred. Risk-control strategies have continuously evolved in response to scientific and technical discoveries and socio-economic developments, as well as from evidence derived from past events. Accordingly, these approaches defined the management of natural hazards in the past century and, as a result, society gradually lost its keen eye for nature's threats and concluded that its dangers remain controllable. Settlements and infrastructures grew further into areas endangered by nature's gravitational forces, but

FIG. 19 A DRAMATIC LANDSCAPE SCENE ON THE MOUNTAINOUS SOUTHERN FLANK OF PIZ CENGALO. THE BONDASCA RIVER CARVED A NARROW GORGE AND FLOWED OVER THE SCREE FIELD INTO THE VILLAGE OF BONDO, SPLITTING IT IN TWO.

1. Walter de Maria, "On the Importance of Natural Disasters," in *An Anthology*, New York: La Monte Young, Jackson Mac Low, 1963.

"I think natural disasters have been looked upon in the wrong way.
Newspapers always say they are bad, a shame.
I like natural disasters and I think that they may be the highest form of art possible to experience.
For one thing they are impersonal.
I don't think art can stand up to nature.
Put the best object you know next to the grand canyon, niagra falls, the red woods.
The big things always win.
Now just think of a flood, forest fire, tornado, earthquake, Typhoon, sandstorm.
Think of the breaking of the Ice jams. Crunch.
If all of the people who go to museums could just feel an earthquake.
Not to mention the sky and the ocean.
But it is in the unpredictable disasters that the highest forms are realized.
They are rare and we should be thankful for them."
WALTER DE MARIA [1]

however much knowledge we evoke, nature remains unpredictable.²

One of the case studies explored in the Robotic Landscapes design research studio is located in Val Bregaglia, one of 150 valleys in the canton of Graubünden. Since 2011, a chain of major tectonic events has profoundly affected this valley as the partial geological collapse of the Piz Cengalo mountain has engendered urgent remediation measures. In August 2017, more than three million cubic meters of fractured rock, mud, and debris collapsed within seconds, causing one of the largest landslides in Switzerland's history and altering the landscape of this otherwise idyllic region. A wall of mud and rocks descended on the community of Bondo, demolishing buildings, blocking key infrastructure, causing the death of a group of hikers, and dividing the village into two disconnected parts. Despite prevention measures in 2015, including the installation of a monitoring and alarm system and the establishment of a prevention project, the overspill basin—formed by monolithic structures of reinforced concrete—failed. The forces of nature partially buried Bondo and caused about forty-one million Swiss Francs in damage.³ Such a major event imposed severe stress on everyday activities and the region's economy, and 1.5 million cubic meters of landslide rock remains in an unstable condition, threatening the Bondasca Valley and its inhabitants. In the village of Bondo, confidence was gradually returning while, on Piz Cengalo mountain, further movements could be detected two years after the catastrophe. Un-

FIG. 20 UNLIKE RECENT DEVELOPMENTS IN BONDO, THE OLD TOWN OF PROMONTOGNO WAS BUILT ON ELEVATED TERRAIN BETWEEN THE BONDASCA AND MERA RIVERS, DEMONSTRATING A LOST TOPOGRAPHIC CRAFT. ALIGNMENT WITH THE POWER OF NATURE SEEMS TO HAVE BEEN FORGOTTEN, CREATING MISPERCEIVED SAFETY WITH PROJECTS SUCH AS THE MASSIVE RETENTION BASIN IN BONDO.

2. BAFU, "Umgang mit Naturgefahren in der Schweiz," in Darbellay, Report of the Federal Council Switzerland, 2016.
3. See www.swissinfo.ch (accessed June 6, 2019).
4. Stéphanie Hegelbach, "Brücke als Balkon," in the Journal of Hochparterre, Zurich, 2020).
5. The lower stream can be defined by the slope of the riverbed where the Gürbe starts to follow the bottom of the valley. See Melanie Salvisberg, *Die unzähmbare Gürbe Überschwemmungen und Hochwasserschutz seit dem 19. Jahrhundert*, Basel: Schwabe Verlag, 2017.

doubtedly, rain will lead to further landslides which could sharply aggravate the precarious situation in Bondo in the coming years.[4]

The most urgent concern the RL design research studio was confronted with in this particular territory was not the strengthening of existing flood protection measures in the region. Instead, it was how to best adapt to chronic natural phenomena and prevent the most damaging impact while creating a new open space in Bondo—namely, a riverscape worth experiencing and living along. To tackle these challenges, numerous opportunities emerged offering topographically resilient structures that invite residents to reconnect with the Bondasca river and its river space, rather than fearing it.

A second case study assessed the Gürbe river that forms a valley between the Bernese Alps and Midlands, characterized by the distinctive lower alpine topography. The 30 km of the Aare tributary has challenging natural conditions and an exceptionally tumultuous construction history. In the 18th century, the river appeared utterly different before the Great Gürbe Correction enabled settlement and economic development in the lower stream.[5] Despite channeling the wild torrent and the construction of 160 dams, flood damage could not be avoided and, today, hydraulic protection projects are still carried out without success. Furthermore, natural hazards in the Gürbe valley have drastically increased in the last century—a combination of rapidly melting snow and long-lasting precipitation have triggered persistent

FIG. 20

landslides. In 1990, heavy rainfall caused a debris flow leading to substantial damage and the destruction of a significant part of the creek structures. After this event, the river's residual sixty barrage steps were restored and reinforced to protect the settlement areas. Looking carefully at the chronic disasters of the last centuries, it can be observed that the number of disruptions caused by natural phenomena will soon increase dramatically. Furthermore, the unstable terrain in the upper watershed damaged the check dams again in 2018, revealing the limits of conventional protective measures on the seemingly uncontainable Gürbe river.[6]

Is it possible to coexist with nature's hazards while preserving the inherent complexity and beauty of a natural landscape? Rehabilitating the current protective infrastructure barriers has become an obsolete endeavor; instead, we must consider the impact of restoring the river space to its natural morphology by removing existing barriers. Resilient topographic structures proved effective in responding to ongoing natural processes and recurring natural calamities. Other than absorbing and redirecting the inherent energies of the natural phenomena occurring on the site, the overall intent behind the structures was to reclaim and form an open and permeable landscape conformation in the territory.

In both cases, local authorities remained confident that the danger of flooding and debris flows was well under control. In 2005 and 2007, the unstable ground of the pre-alpine and alpine regions revealed

FIG. 21 WHETHER A NATURAL EVENT IS A DISASTER OR NOT DEPENDS ULTIMATELY ON ITS LOCATION AND PROXIMITY TO HUMAN-MADE ARTIFACTS. VISIBLE HERE IS THE CROSSING OF A LOCAL ROAD OVER THE CHANNELED GÜRBE RIVER.

6. Melanie Salvisberg, "Der Hochwasserschutz an der Gürbe, Eine Herausforderung für Generationen (1855–2010)," in *Wirtschafts-, Sozial- und Umweltgeschichte*, 7, Open Access, Basel: Schwabe Verlag, 2017.

the inadequacy of the implemented measures and the urgent need for strategic approaches to cope with the emerging environmental challenges of changing climatic conditions. With this urgent need to mitigate the effects of hydro-meteorological disasters, the Swiss authorities—after years of planning—increased investments in protective structures to cope with the dangers of floods, avalanches, forest fires, landslides, and rockfalls. While proposed projects solve local challenges, they do not take large-scale morpho-dynamic processes and the activation of these unique landscapes into account. The Robotic Landscapes design research studio encourages a reflection on beauty alongside these events' relative magnitudes in design strategies. Our work highlights how the mitigation of natural disasters is a permanent task and can be interrelated in a design process as a catalyst for feasible topographic structures to incentivize a conscious and critical understanding of the given realities. The ultimate objective in dealing with these phenomena is to ensure sufficient protection, but also to conceive multifaceted and sustainable living spaces. A consideration of the inherent potential of natural dynamics with topographic intricacies includes a great and unique opportunity to reimagine the design of landscapes in times of climate change.

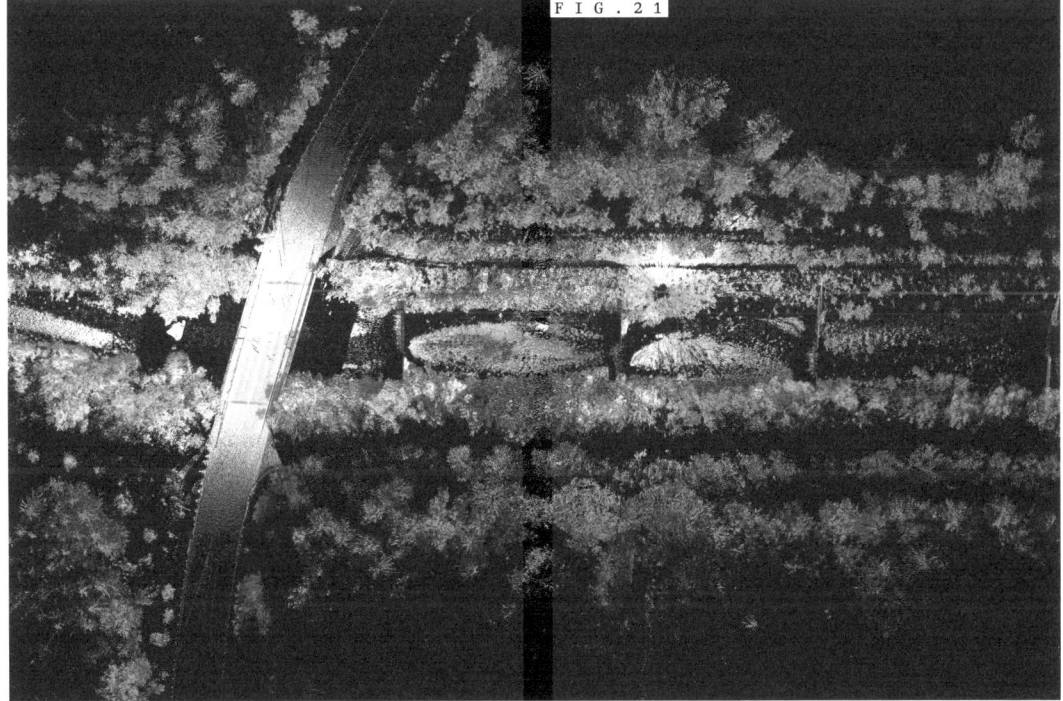

FIG. 21

FIG. 22 A DIGITAL RECORDING OF THE BONDASCA RIVERBED SCATTERED WITH GRAVEL AND BOULDERS AFTER THE DEBRIS FLOW EVENTS OF 2017.

FIG. 23 THIS IMAGE SHOWS THE TRANSITIONAL ZONE OF THE GÜRBE RIVER, SITUATED BETWEEN A STEEP GORGE AND AN ALMOST HORIZONTAL VALLEY FLOOR. DISTINCTLY EXPOSED IS THE CANALIZED RIVER WITH ITS MANY RETAINING DAMS.

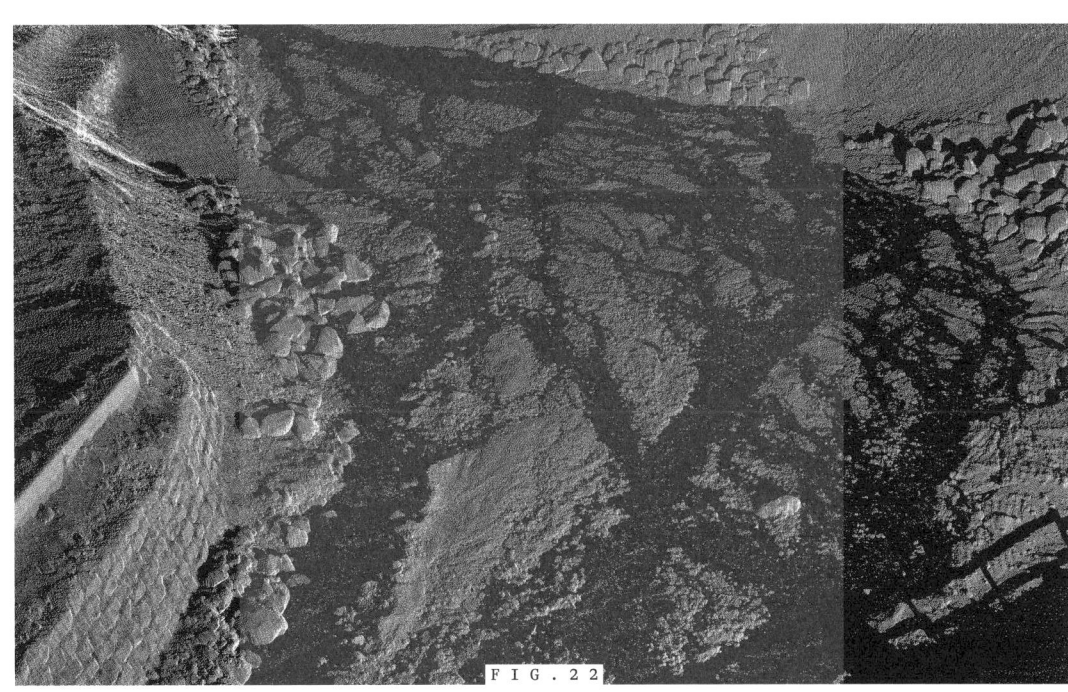
FIG.22

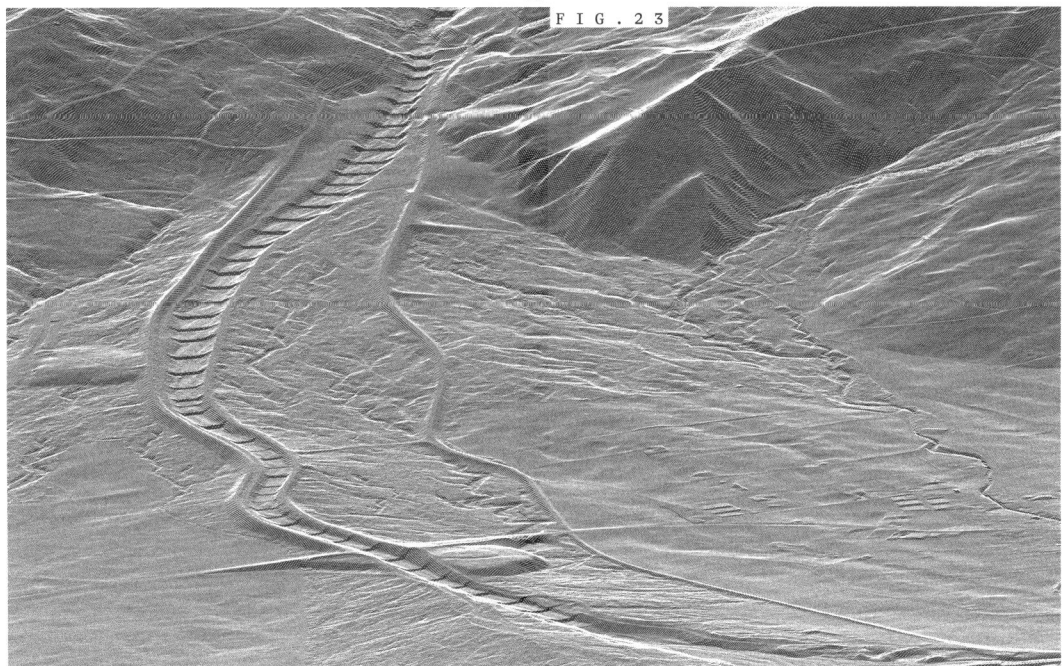
FIG.23

FIG.24 THE INCONCEIVABLE REALITY AND MAGNITUDE OF THE EVENTS REVEAL THE POWER OF THE NATURAL PHENOMENA THAT DESCENDED ON THE TOWN OF BONDO IN THE COURSE OF JUST A FEW DAYS.

FIG. 25 IN THE SCOPE OF A DISASTER, HUMAN VULNERABILITY AND A DISPOSITION TO RESIST TAKE ON A PROFOUND SIGNIFICANCE IN THE RECOVERY OF THE LANDSCAPE.

FIG. 26 THIS POINT CLOUD REPRESENTATION PORTRAYS THE PARADOX BETWEEN PICTORIAL BEAUTY AND THE TREMENDOUS EFFORTS UNDERTAKEN TO ESTABLISH THIS RIVERSCAPE.

OPERATION

The connection between the digital and the natural can be captured through the use of sophisticated sensing devices, leading to successive design chains where the final result is not stagnant, but part of a continuous process of formation. Instead of applying complete control over the outcome, designers must now learn to interact with digital media to help better direct the flow of forces and of matter in the landscape. These become part of a computational system that acts as an environmental mediator with its own internal logic and constraints. Drawings, models, maps, and physical experiments from this design research process help exhibit the inherent relationship that prevails in the framing, forming, and finding methods. Framing is the act of gathering principles that affect the material realm under study, including robotic platforms, natural phenomena, as well as design objectives. Forming responds to the design action itself, exploring and proposing solutions by using dynamic geometries and procedures over time. Finally, finding apprehends the outcomes of a design process by using, for example, computational methods capable of simulating material behaviors and testing their efficiency. These three principles together redefine the way designers can think, act, and engage in contemporary landscape architecture.

Given the ongoing progression in natural and robotic systems, how can we best draft designs that accommodate continual adaptation? It is clear that design practice must break from its habits and evolve as well. Through the recursive application of framing, forming, and finding, a feedback loop can be engaged with the current status of the landscape, informing and guiding the architect's response. Any moment in the process can constitute the basis for the next design operation, and this can be repeated indefinitely. The opposite holds true as well: any attempt towards the elaboration of a "final" project is delusive since it will always be the result of an ongoing continuum. *Operation* is therefore not concerned with prescribing geometries; on the contrary it is about discovering form as it continually unfolds.

The actions of surveying, prototyping, processing, manipulating, simulating, evolving, and maintaining depict how a design scheme can progress from early topological investigations into active topographic maintenance scenarios. By showcasing this research-in-the-making, we aim to rethink how we best portray terrain as a topological constant in continual transformation. It demonstrates a strong problem-solving framework that continuously questions and verifies itself through the fundamental parameters of a given proposal. By crafting these yet-unfinished topographic concepts resulting from countless design iterations and feedback loops, we can now openly question the legitimacy of the linear approach to design and material culture that we have inherited and practiced to date.

To conceptualize the complexity of a terrain in terms of its morphological and topological elements, a site is carefully approached and measured in its raw state. Focused on a comprehensive analysis of existing forms and processes in nature like soil, water, flora, space, and time, these initial investigations offer the possibility to perceive a dynamic landscape both with our senses and our sensors to uncover their specificity.

The landscape body, as seen from a human perspective, gains a dimension and materiality that is captured and brought back to the design research studio. On contact with the terrain to be developed, one classifies its dimensions, its temporal aspects, and its visual splendor. By drawing diagrams or maps, one makes a logical two-dimensional image that rarely looks like the landscape it stands for. These logical pictures abstractly represent the site, subsequently becoming the starting point for drafting tangible architectural expressions. Here, the landscape experience is creatively translated into articulate reflections on the tangible problems of a site, helping to create a critical understanding and interpretation of its contents.

Surveying is the act of recording the area and features of a piece of land, and the results of these temporary displays may span all media. This analytical reflection exploits the primary source materials by highlighting the sites of study, framing their foundations, and acknowledging the creative processes that occur in their conceptual representation.

"Should we hope to govern nature or should we give it carte blanche, allowing it to take the blame for its most biased and murderous manifestations?"
CARLO MOLTENI

"Morphing can suggest many things, but here it describes a change in time, evolution, destruction, and reconstruction. It represents cyclical creation through decay."
KELLY MENG

"Are the effects of floods, landslides, avalanches, and debris flows necessarily disastrous? How can we allow natural happenings to take their course, guiding nature using robotic processes to achieve a design goal?"
SAKIKO NODA

SURVEYING

CONVEYING SPATIAL AND MATERIAL DIMENSIONS
TRANSMITTED BY AN IMAGE IS THE MOST IMPORTANT POINT
OF INTEREST, RATHER THAN THE DEPICTED SCENE.

F R A M I N G

107–119

F I N D I N G   F O R M I N G

IDENTIFYING TOPOGRAPHIC PECULIARITIES IN THE FIELD UNDERSCORES ITS ONGOING METAMORPHOSIS. THE ACCUMULATION OF EXISTING, TRANSFORMED, AND NEWLY ARRIVED MATERIALS OVER TIME RESULTS IN A COMPLEX MINERAL SYSTEM SUSCEPTIBLE TO REALIGNMENT.

DESCRIBING MINERAL ENTROPY IN NATURAL AND HUMAN-MADE MATERIAL ACCUMULATIONS IN A POST-NATURAL DISASTER AREA REVEALS A SCALE OR ARRANGEMENT OF MATERIALS IN PREVIOUSLY IMPERCEPTIBLE DIMENSIONS.

FORMING FINDING FRAMING

108

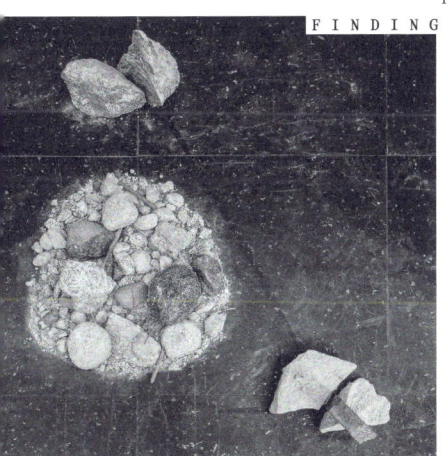

FRAMING
FINDING FORMING

FORMING FINDING
FRAMING

110

FRAMING

111

FORMING

INDING FORMING

FRAM

FINDING

FRAMING

113

FINDING FORMIN

MING

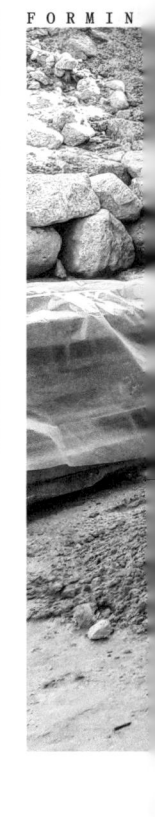

FORMING FINDING
FRAMING
114

FRAMING
FINDING FORMING

PROTOTYPING

"A terrain is like an open book; it gives an enormous amount of information to anyone who cares to read it."
KELLY MENG

"Working with our hands gives us a greater appreciation for the precision and the repeatability of robots. It also allows us to better understand the distinction between human and machine—within the same sandbox, different possibilities are opened for the two different actors."
LEE LIP JIANG

"How exactly can a river be thought of topographically, and how can one think of surface as both a functional tool and an expressive instrument? How exactly does the robotic process relate to such questions of design?"
LORIN WIEDERMEIER

In response to the constraints and pressures imposed on a site, new topographies are proposed based on the specific means of production. Here, landscape structures are formed that fulfill four criteria. One, they balance the cut and fill of locally available granular material; two, they form naturally stable aggregations; three, they can be produced using robotic earthmoving procedures; and four, they create a dynamic equilibrium within the existing processes on site.

In sketching and manual sandbox modeling, an intuitive sculpting process bridges the gap between the designer and the physical reality of a landscape. This starting point is deliberately free of formal constraints and explores the topography's potential to steer natural processes. In an iterative fashion, models acquire their specific geometry through digital terrain modeling, ready for testing in large-scale landscapes.

Prototyping is the act of making basic models to explore the design space of a particular system or architecture. From a designer's intuition to a performative topology for dynamic earthmoving strategies, the drafting process phrases sophisticated topographic modulations.

EXPLORING THE TOPOGRAPHIC CONTINUITY OF SIMPLE INTERSECTING PLANES IN A THREE-DIMENSIONAL MODEL YIELDS MULTIFACETED COMPLEXITY.

EVALUATING THE FORMAL STRATEGY IN A DIGITAL TERRAIN MODEL BY ILLUSTRATING THE WATER DRAINAGE PATTERN DISCLOSES A MYRIAD INTRICATE LANDFORMS.

FORMING FINDING
121–131
FRAMING

INCORPORATING A DELIBERATE AND INTELLIGIBLE TOPOGRAPHICAL LOGIC—WHERE AN AWARENESS OF THE GIVEN REALITY EMERGES—IS IMPLICIT IN THE INITIAL STRATEGIC-THINKING PROCESS APPLIED TO A DEBRIS FLOW MITIGATION STRATEGY.

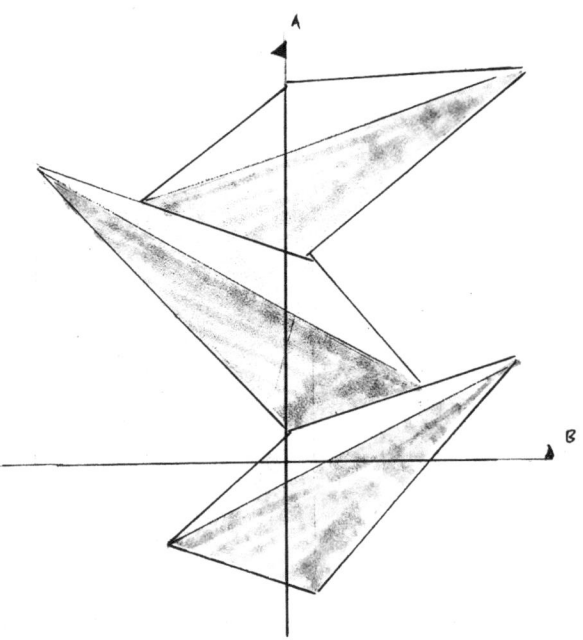

FRAMING
122
FINDING FORMING

FORMING FINDING
FRAMING

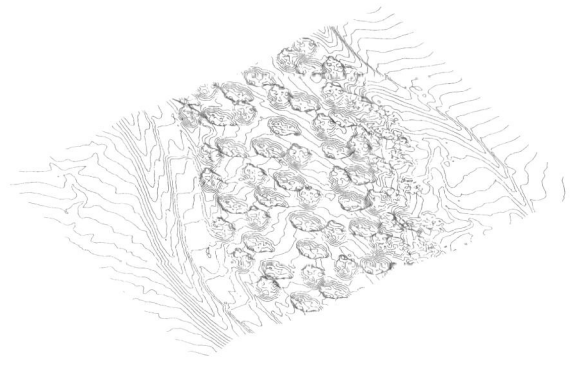
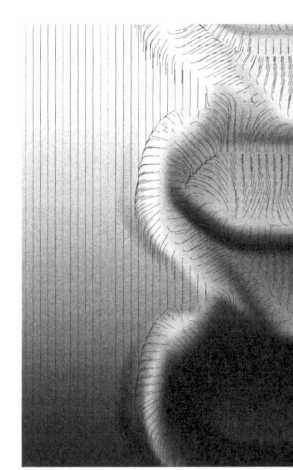

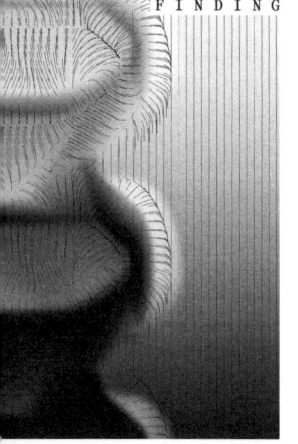

FRAMING
124
FINDING FORMING

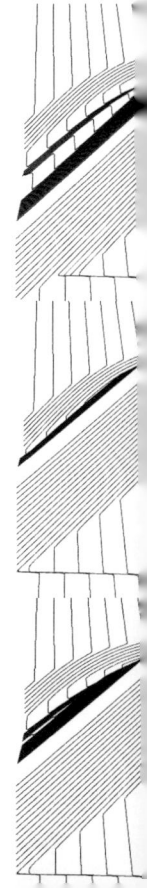

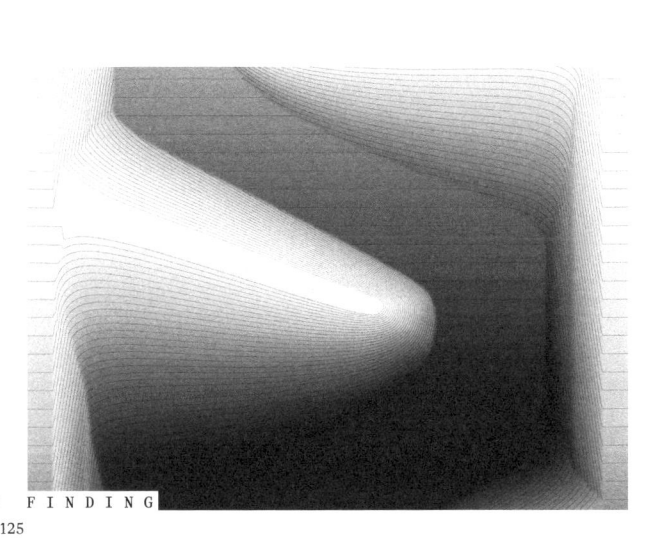
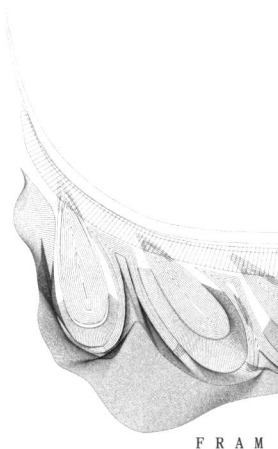

ORMING FINDING   FRAM

125

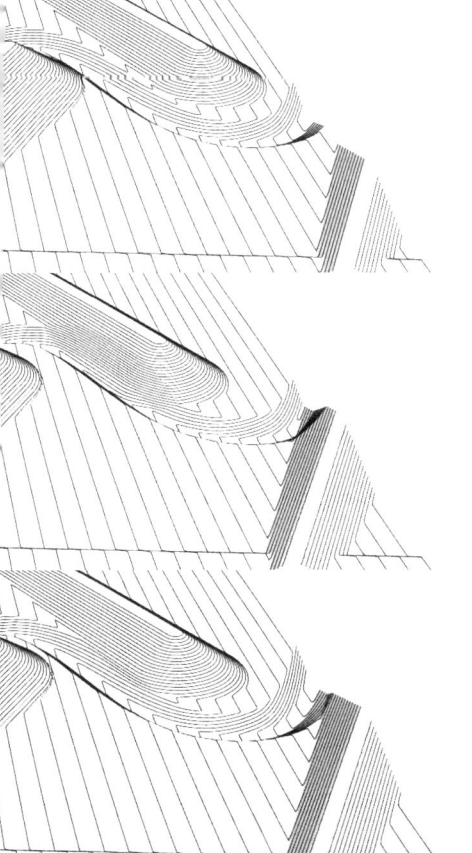

FRAMING   FINDING

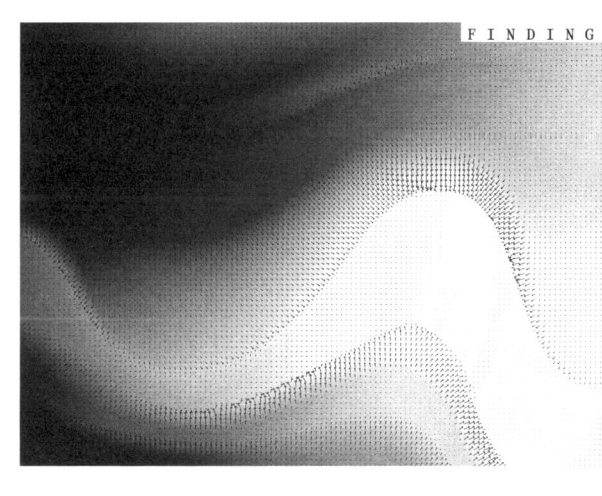

ING   FORMING FINDIN
FORMING   127
                     FRAMING

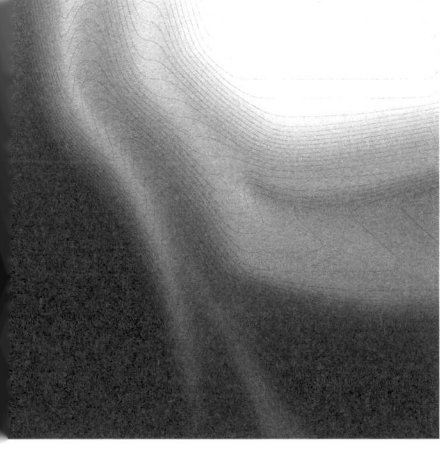

128

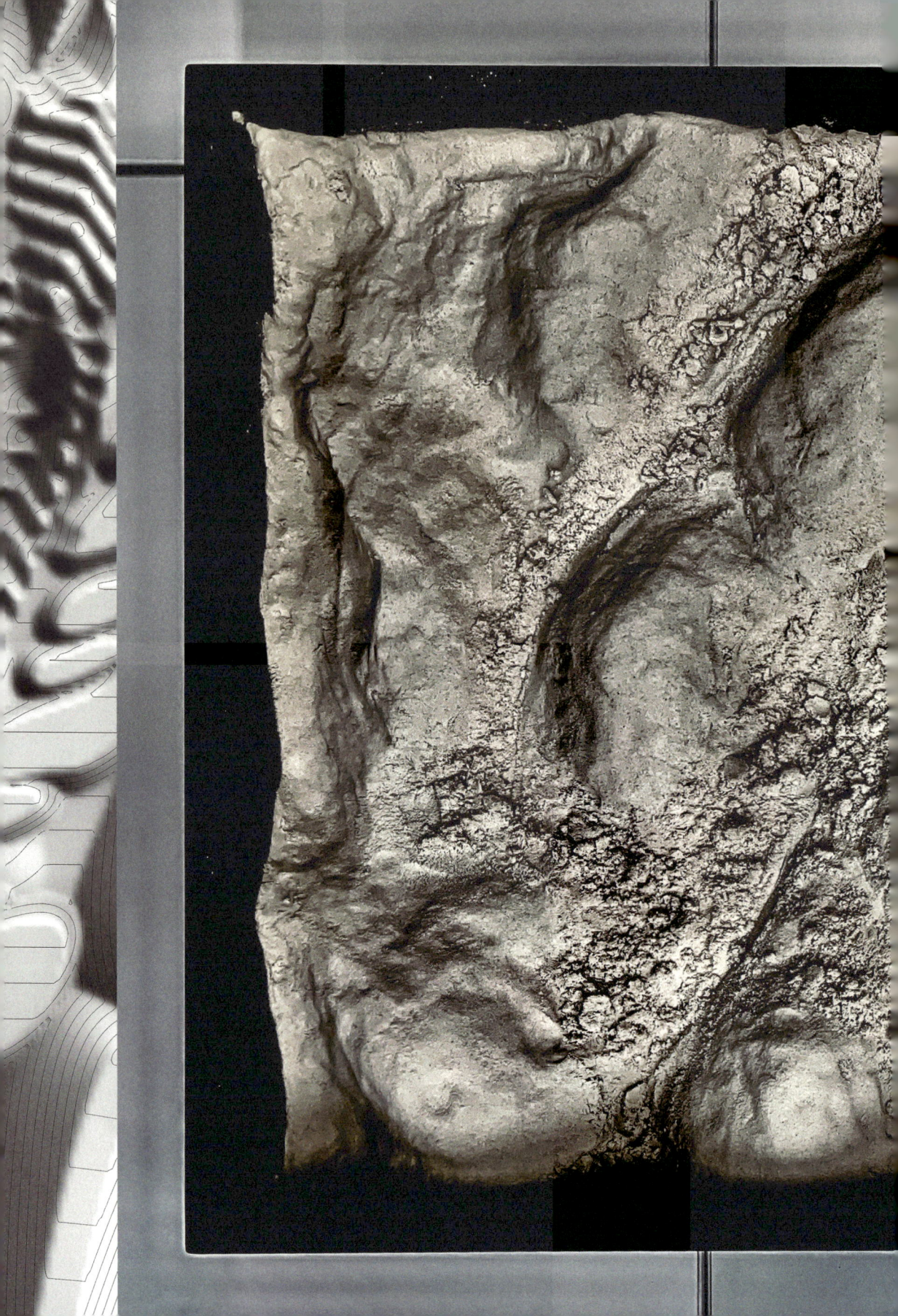

Leo und Hannah

Robotic movements are not predefined, they are the result of a specific pre-existing condition in an 80 × 120 cm box. The sandbox model's topography is first scanned in three dimensions, allowing for digital examination before a robotic movement is encoded through analysis, data-mapping, and translation, from derivatives like elevation, slope, and the orientation of the scanned topography.

No two designs will ever come out the same because the robotic movement is inherently linked to the sandbox, and the formation of the granular material behaves differently over subsequent iterations. This indeterminacy in cyber-physical systems is encouraged to inspire unexpected topographies, performances, and pleasing aesthetics in granular material systems. Rather than simply being about control, these experiments search for moments of performance where the integration of materials and digital processes generates a new landscape choreography.

Processing is the act of performing a series of computational operations to encode a robotic movement. Computational procedures transform the sandbox over time in an abstract and prototypical manner, revealing unexpected realities and conveying inconceivable realities beyond human imagination.

"How can complex problems and sequences be translated into computable language? How can a question of meaning be turned into one of logic?"
LORIN WIEDERMEIER

"Programs, codes, language—they are merely ones and zeros transmitted to the main control system, manifesting themselves in a series of outputs. In the case of the formidable Menzi Muck, this is manifested as calculated physical motion. Robots are essentially anthropomorphized entities, and in constructing intuitive artificial movement, there is a need to deconstruct what it means to move intuitively."
TOH WEI WEI

"First, a robot will scan an existing topography, extracting data to be analyzed. The next step is made by the designer who draws the first level of curves onto the territory in relation to the data received by the robot. These will define the zones for the materials that form slopes. Subsequently, these curves will be projected by the robot onto the existing terrain. The distances between the points will be compared, creating a connection between the curves and the topography. The robot will then calculate the amount of material to be deposited across the topography and how these curves can be achieved, varying the speed with which material is deposited."
CARLO MONTENI

PROCESSING

ENCODING THE FOUND TOPOGRAPHY OF THE SANDBOX USING A
3D SCANNER OPENS A WINDOW TO THE OBSERVABLE SURFACE
STRUCTURES IN THEIR ENTIRETY.

FRAMING

133-149

FINDING   FORMING

TRANSFORMING THE MATERIAL IN THE SANDBOX INTO A
PERFORMATIVE TOPOGRAPHY DRIVEN BY HIGHLY RIGOROUS
ROBOTIC FORMATION PROCESSES.

CONVERTING THE DETECTED TOPOGRAPHY THROUGH DIGITAL
ALGORITHMS INTO A SEQUENCE OF DYNAMIC OPERATIONS
THAT ARE ENCODED FOR ROBOTIC OPERATION, PROVIDING
IT WITH AN INTERFACE MODEL TO CORRELATE WITH TIME,
PLACE, OR A SPECIFIC GEOMETRY.

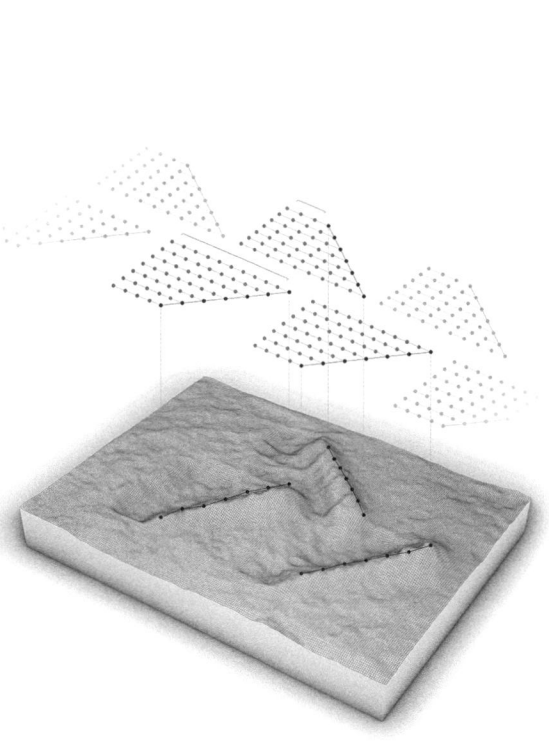

FORMING FINDING
FRAMING

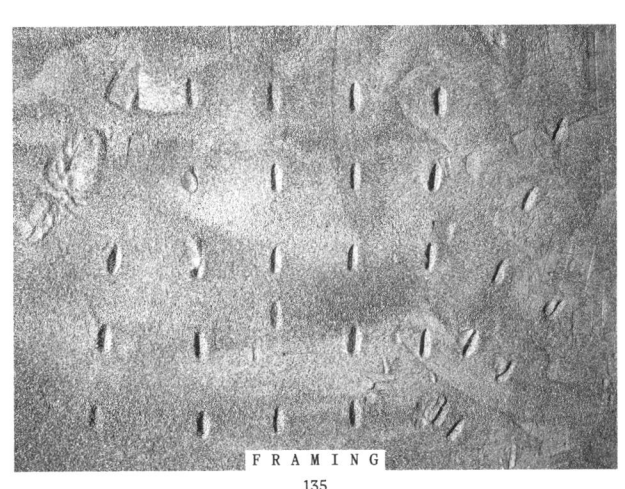

FRAMING
FINDING FORMING

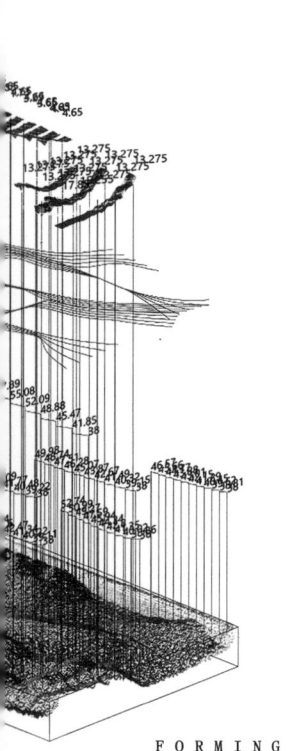
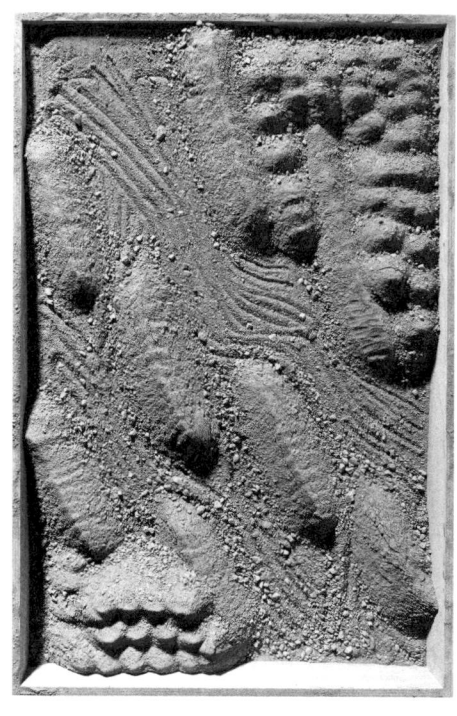
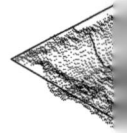

FORMING FINDING
FRAMING

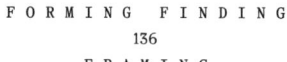
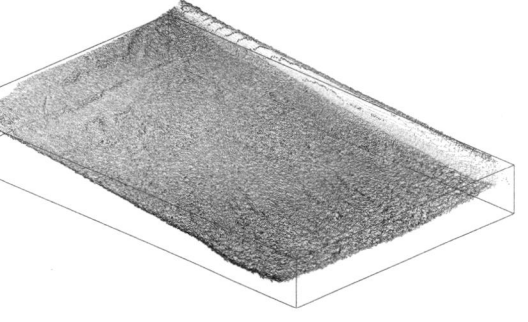

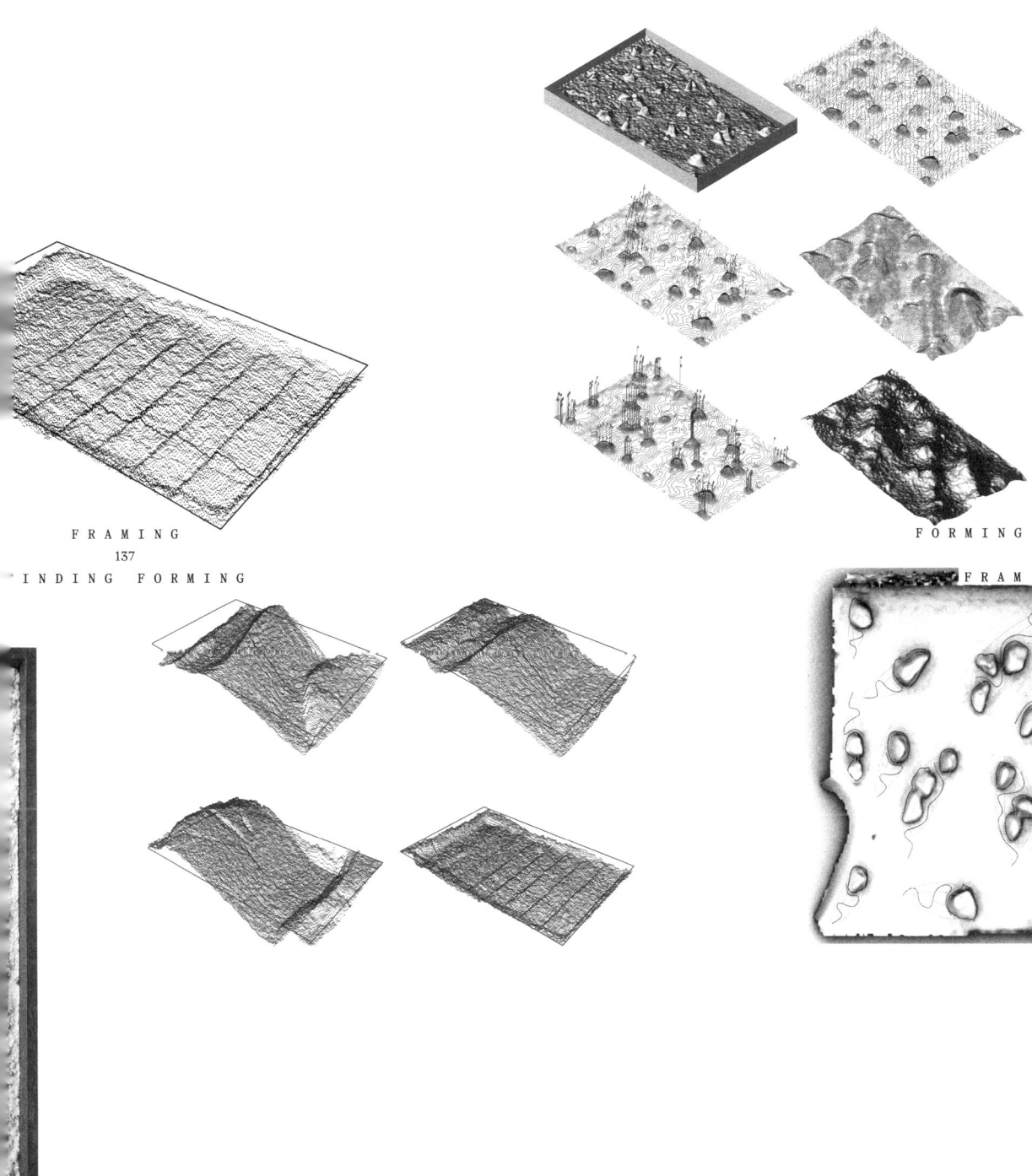

FRAMING
137
INDING FORMING
FORMING
FRAM

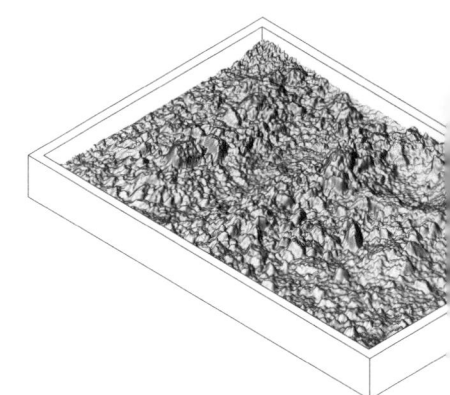

FINDING

FRAMING

ING

FINDING FORMING

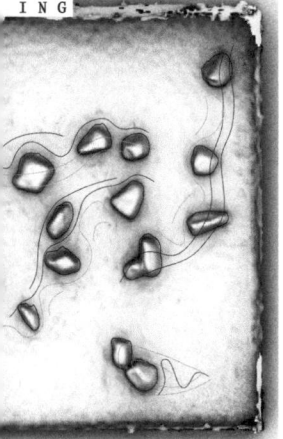

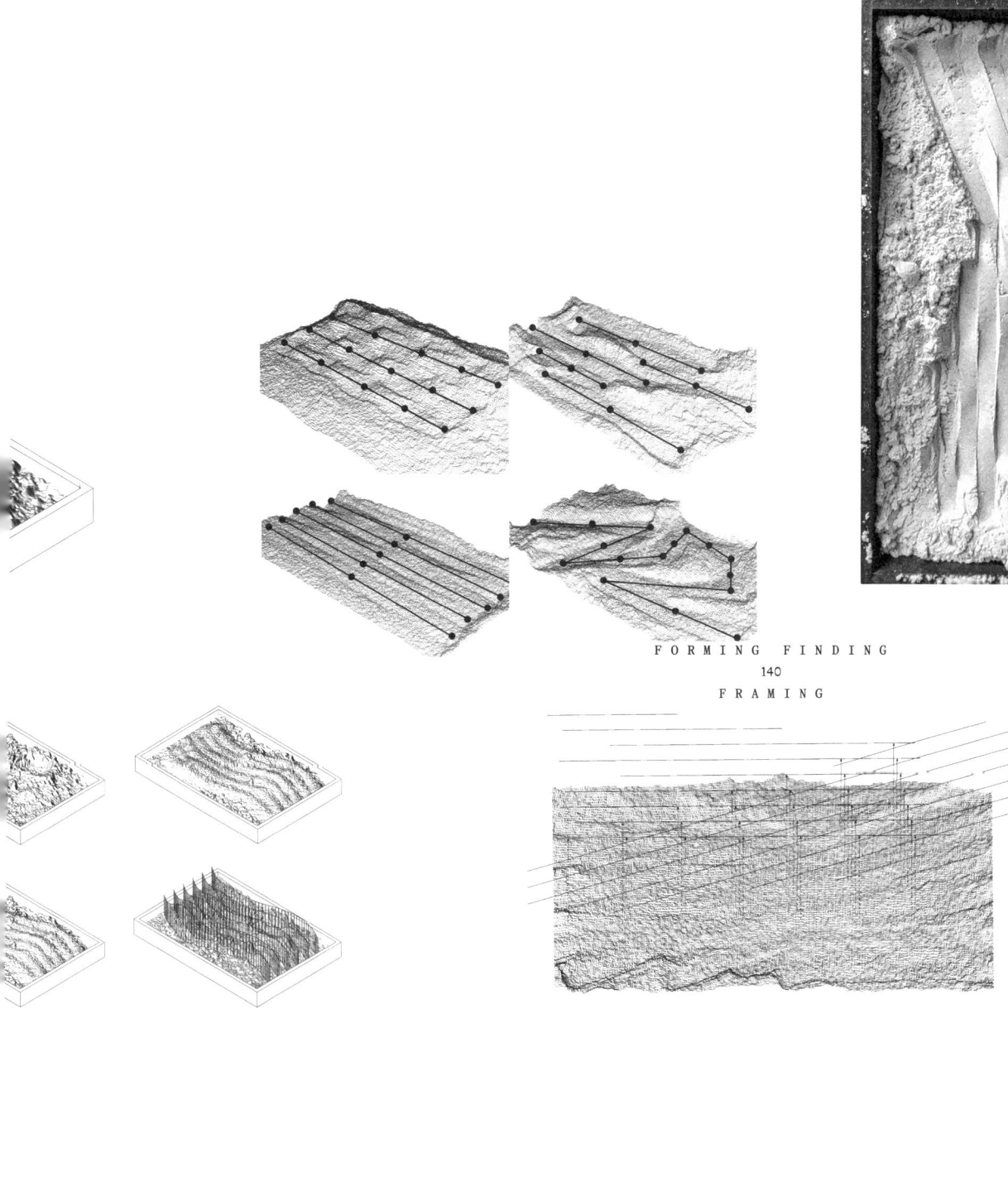

FORMING FINDING
140
FRAMING

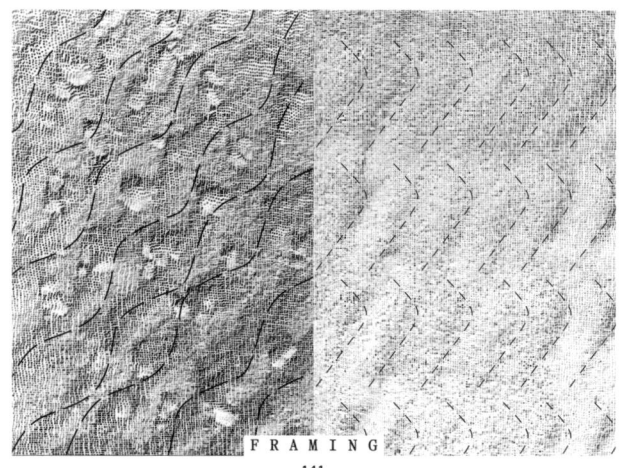

FRAMING
141
FINDING FORMING

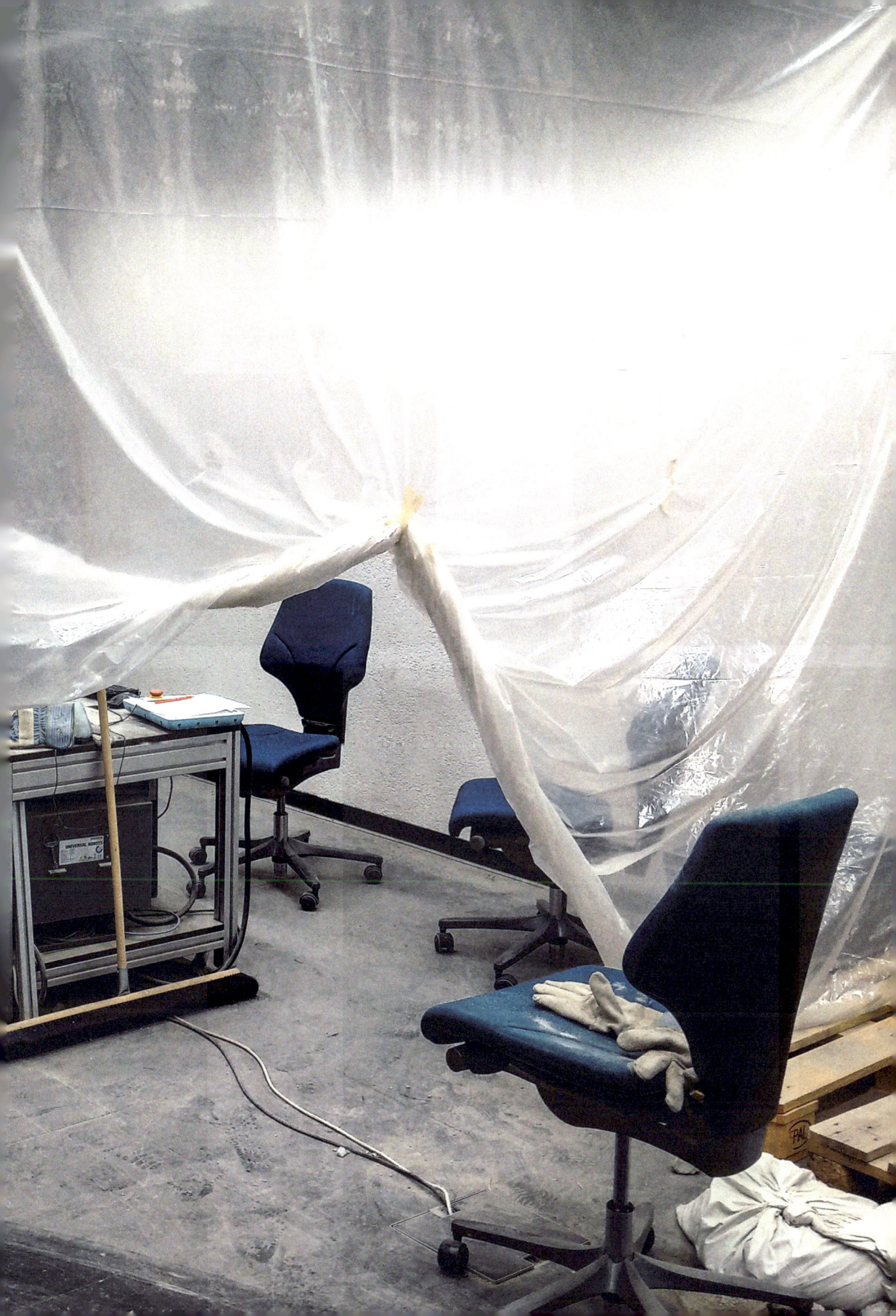

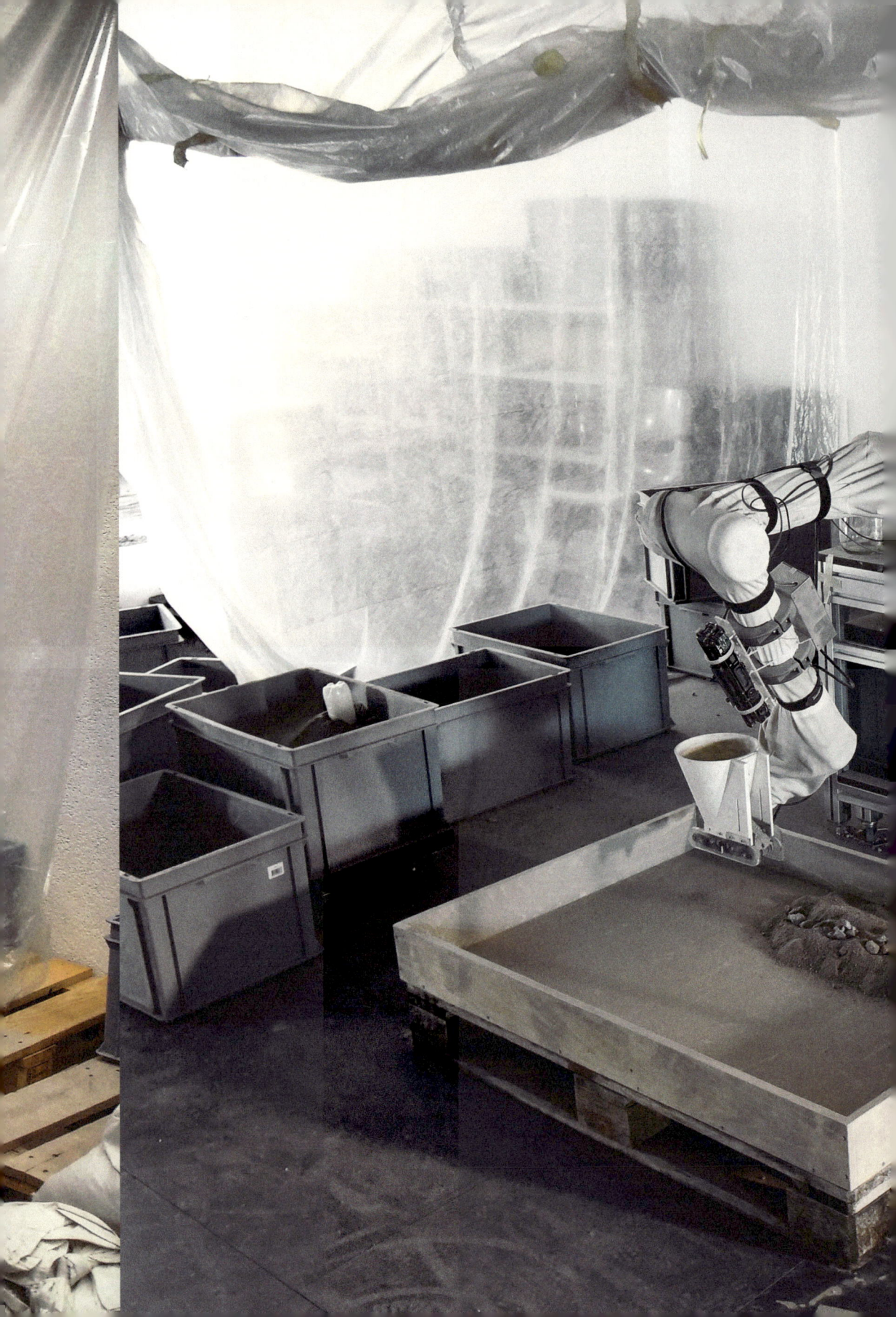

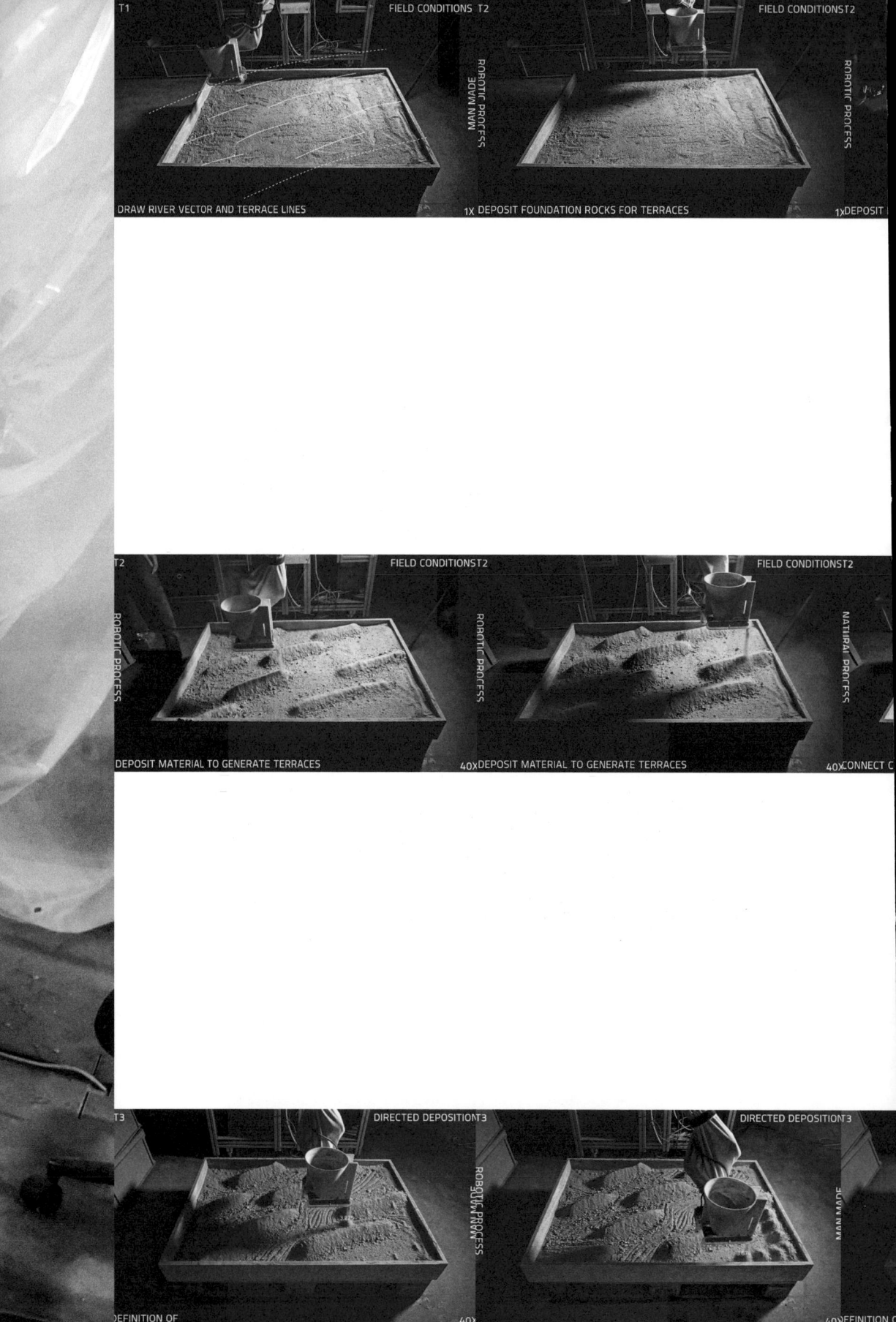

MANIPULATING

"The interaction between human, robot, and nature will always be an open subject, but this project aims to find an equilibrium where each of these domains can freely flourish without crossing the boundaries of the other."
KELLY MENG

"Feedback loops consisting of an analysis of a terrain, and a reaction to it, are used to integrate robotic processes in a project. Depending on the slope or size of the material (boulders, rocks, sand), the riverbed texture will be developed into a pattern to be followed by the machine."
KELLY MENG

*"Dust and sand everywhere.
Dust on my clothes
dust in my pockets
dust in my hair.
Sand on my floor
sand in my coffee
sand on my screen
sand in my keyboard
sand on my chair.
Crunching under my shoe
crunching under my laptop
between mouse and table
between my teeth."*
JONAS HELDEMANN

By equipping a robotic platform with an end-effector, a 3D scanner, and force feedback, it becomes possible to autonomously adapt to the many geometries found in the sandbox, making any interaction specific and unique. By dedicating various end-effectors for different cycles, each subsequent manipulation takes part in an increasingly complex configuration in the sand.

While real-world earthmoving equipment is designated for specialized tasks, robotic manipulation offers a combination of operations that were previously unthinkable. This has an influence on the design of the end-effector, where dumping and compression cycles can be combined in a single movement. This strategy uses simple tools to accomplish complex tasks to create surprising effects, leveraging computation while minimizing mechanical complexity.

Manipulation is the act of handling or moving material using a robot's end-effector, whose tools directly interact with the material at hand using shifting, compressing, spreading, digging, or dumping cycles. The end-effector offers a succession of intricate resolutions, where natural formation and robotic manipulation in granular material play out between deliberate organization and entropy.

THE SCULPTING OF GRANULAR MATERIALS ORIGINATES FROM
THE ROBOTIC MOVEMENT AND THE SPECIFIC SHAPE OR
ACTION OF THE END-EFFECTOR. THE RANGE OF POSSIBLE
SCENARIOS IN THE RAW MATERIAL THAT PLAY OUT IN THE
BOX IS INEXHAUSTIBLE.

CHALLENGING THE EXPECTATIONS SET BY THE CONCEPT,
EACH MODEL EXHIBITS A CREATIVE, DISTINCT, AND
UNEXPECTED FORMAL EXPRESSION.

FORMING FINDING

151–163

FRAMING

APPLYING ITERATIONS OF THE DYNAMIC OPERATING CYCLE
OF THE END-EFFECTOR LEADS TO UNIQUE TOPOGRAPHIES.
AN INHERENT DESIGN PREMISE IS ADDRESSED
THAT CONTEXTUALIZES THE MOVEMENT'S LOCATION
IN THE SANDBOX.

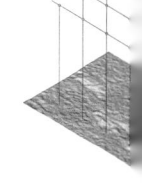

FRAMING
152
FINDING FORMING

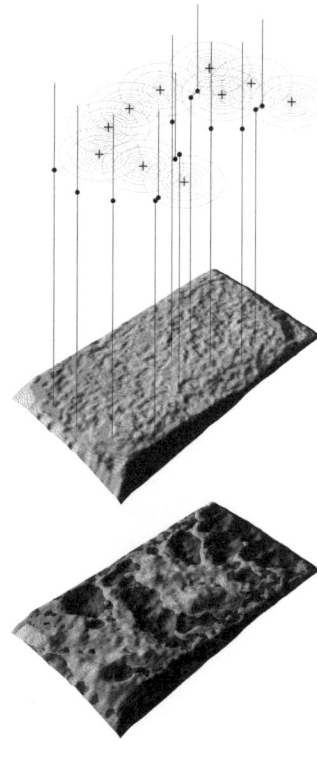

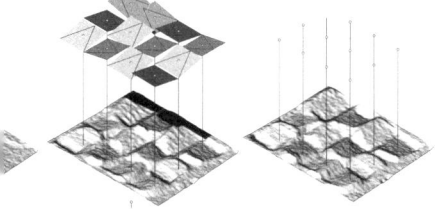

FORMING FINDING
FRAMING

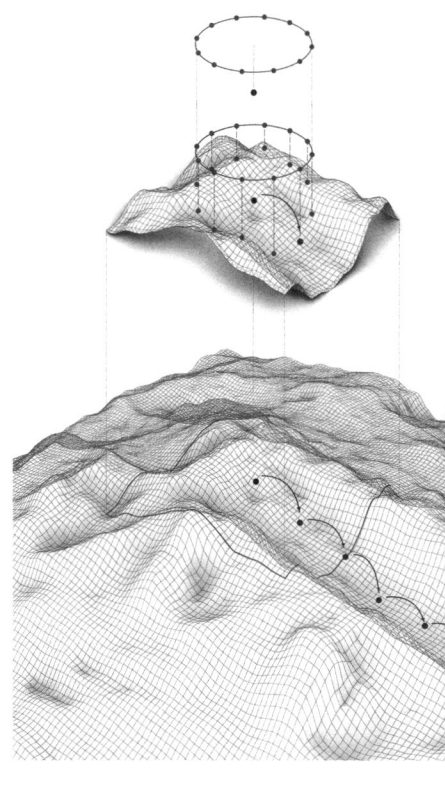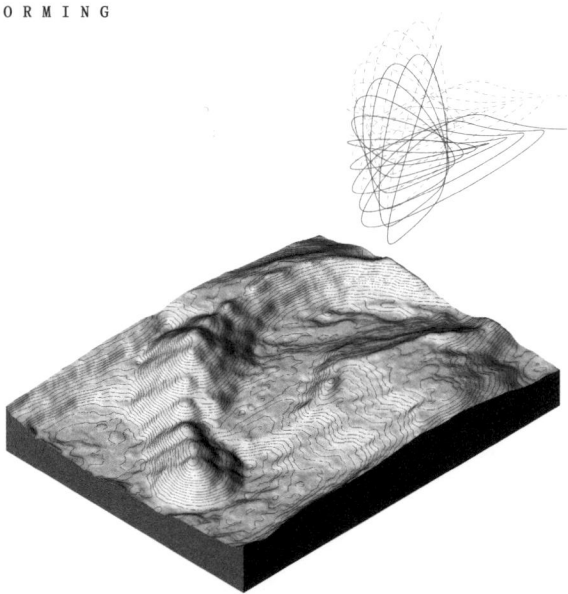

FRAMING
154
FINDING FORMING

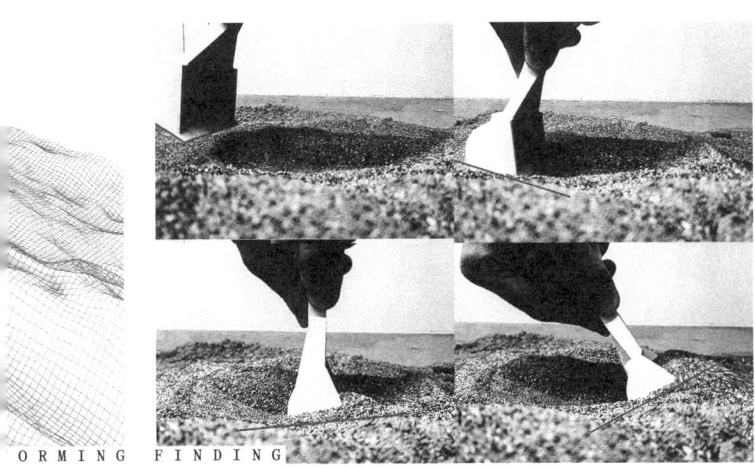

ORMING FINDING　　　　　　　　　　　　　　　　　　　　　　　　F R A M

155

FRAMING　　　　　　　　　　　　　　　　　　　　　FINDING

ING

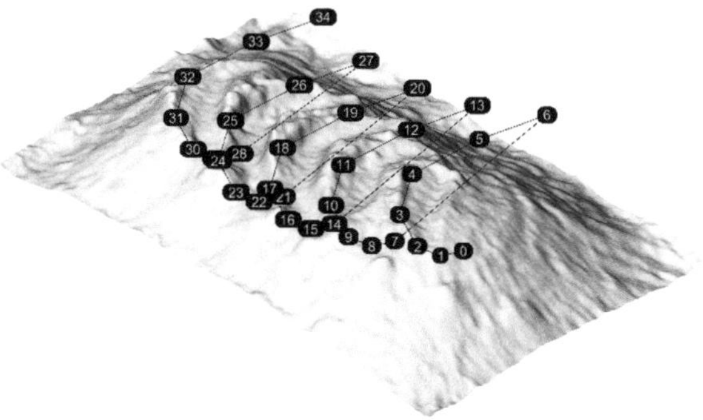

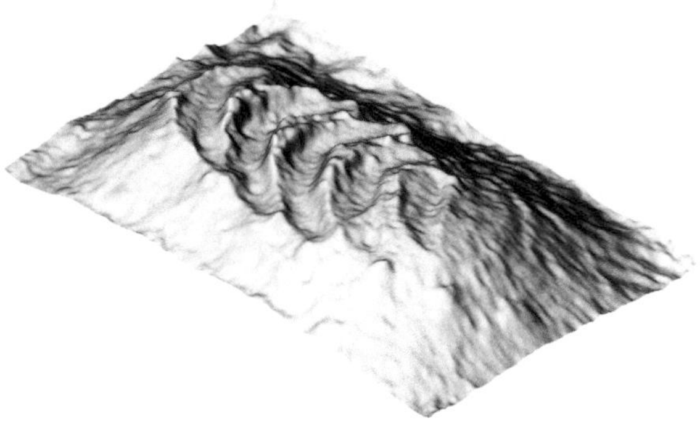

FORMING

FORMING FINDING

FRAMING

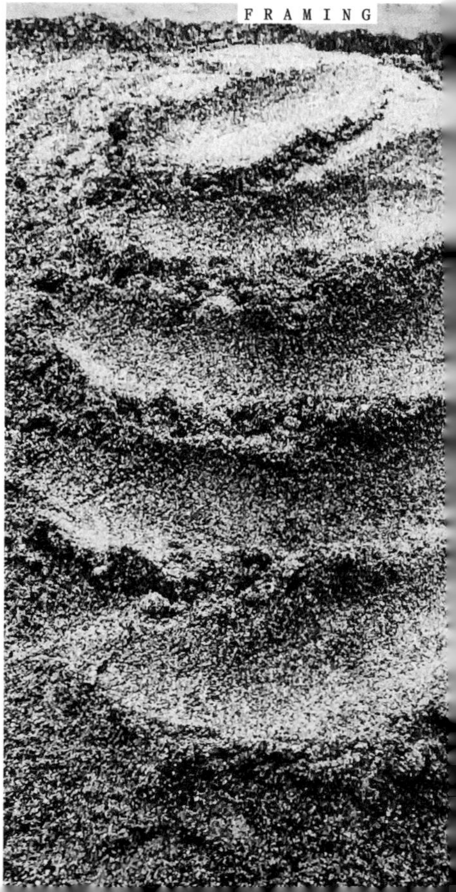

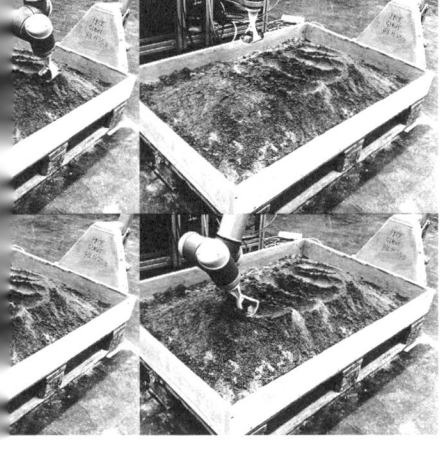
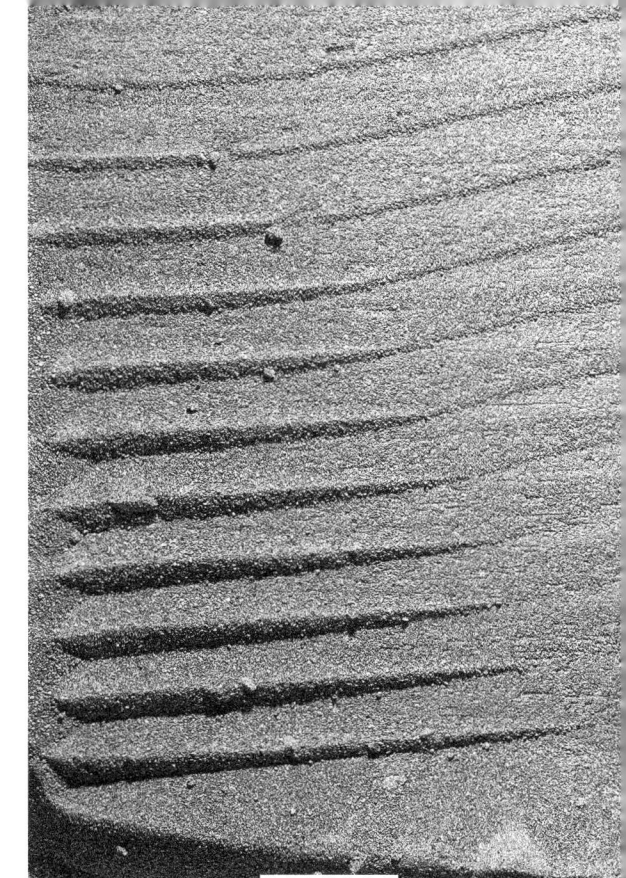
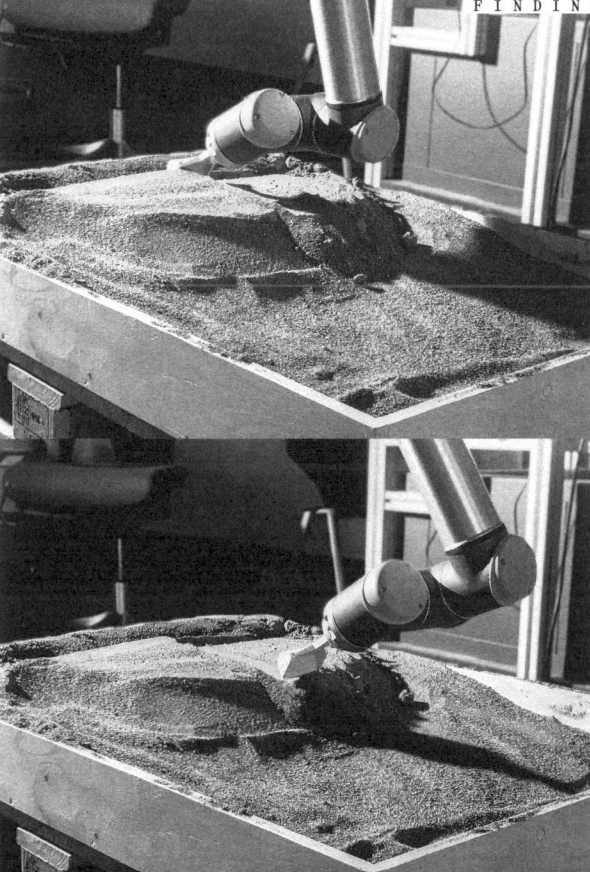
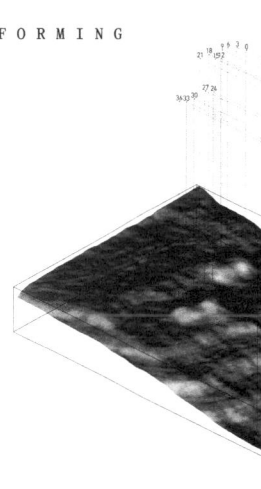

FRAMING
FINDING FORMING

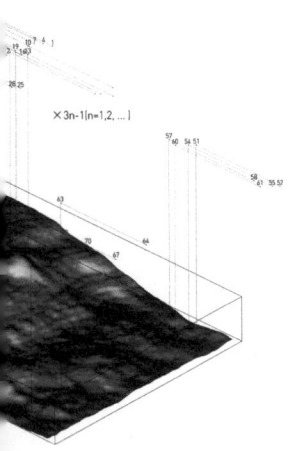

FORMING FINDING
FRAMING

FRAMING
160
FINDING FORMING

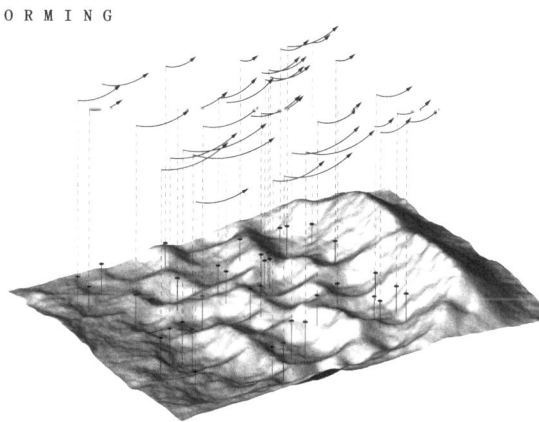

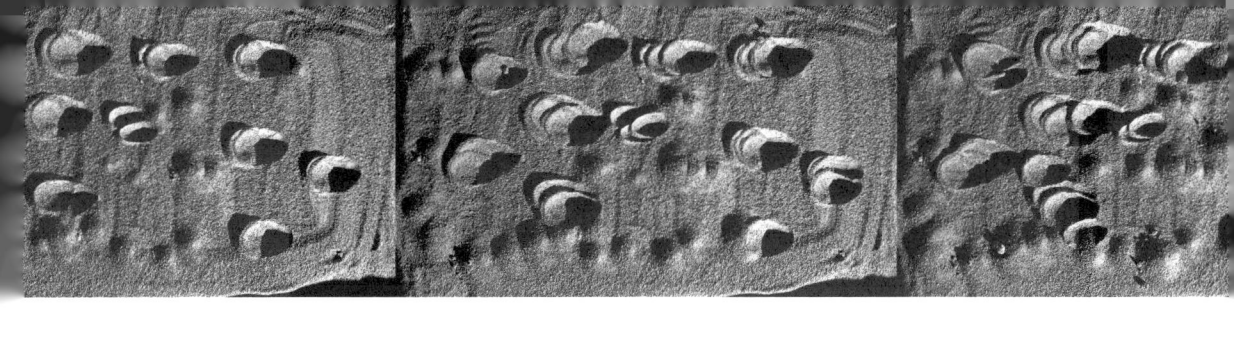

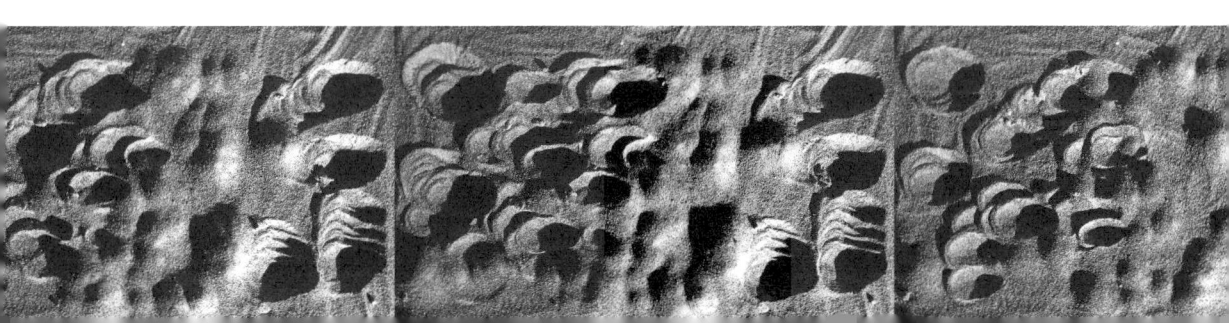

The ways debris flows and landslides perform on various topographic designs is tested using simulation. The flow of mass movement processes on proposed topographic schemes reveals their performance. By iterating erosion and deposition spanning large time frames, a continuous buildup of the landscape is modeled and visualized. After each simulation, robotic modeling strategies alter and steer the way these natural processes behave, preparing a responsive terrain structure that is ready to safely receive imminent mass movement events.

A design does not consist of a single vision but is instead a chain of procedures that unfolds over time. While it is possible to forecast likely scenarios using simulation, it is virtually impossible to anticipate natural events that lie far in the future. As such, a computational approach that dynamically accommodates change is vital. Starting from the existing situation (t=0), designs play out natural processes and robotic earthmoving procedures over time.

Simulating is the act of imitating a continuous physical process using digital calculation. Here, natural processes such as erosion, transportation, and sedimentation are dynamically modeled on a terrain. In the design process, simulation methods lead to the emergence of new landscape structures that are dynamically built up through iterative mass movement simulations.

"What can a simulation reveal and how can it be a tool for design? Simulations clearly reveal whether or not a system performs as desired, showing that the movement of a mass and a topography create a delicate equilibrium where slight changes to the topography can have massive consequences."
LORIN WIEDERMEIER

"Debris flows are a question of volumes— a simple mathematical balancing act. What form does 400,000 cubic meters take? Is it a 1 × 632 × 632 meter volume? Is it a 2 × 447 × 447 meter volume? Is it 3 × 335,000 × 335 meter volume?"
LEE   LIP   JIANG

"The first intervention is balanced between cut and fill. This is a challenging part of a design where the amount of excavated material creating the retention basin has to be placed somewhere. After that, more material is deposited on site through debris flows, and this excess material also needs to be placed. The project in itself is about finding an equilibrium between 'le triangle infernale' of nature, architecture, and robot."
KELLY   MENG

SIMULATING

BALANCING THE MATERIAL VOLUME IS A PREREQUISITE IN COMMITTING TO A SUSTAINABLE AND RESILIENT LANDSCAPE. DYNAMIC EARTH MOVEMENTS ARE MODELED TO DETERMINE WHERE EXCAVATION (BLUE) AND DEPOSITION (RED) IS REQUIRED.

FRAMING

165–177

FINDING   FORMING

EVALUATING A TOPOGRAPHY'S PERFORMANCE USING MASS MOVEMENT SIMULATION. EROSION AND SEDIMENTATION PROCESSES ARE EMBEDDED IN THE DIGITAL TERRAIN MODEL TO REFLECT THE TEMPORAL MATERIAL VOLUME BALANCE.

ARISING FROM NATURAL AND ROBOTIC PROCESSES, A NEW LANDFORM EMERGES BY SHIFTING THE FOUND GRANULAR MATERIAL INTO A PERFORMATIVE TOPOGRAPHY. HERE, CNC MODELS SERVE AS A PHYSICAL MEDIUM TO INTERPRET THE PROJECT'S TECTONIC EXPRESSION.

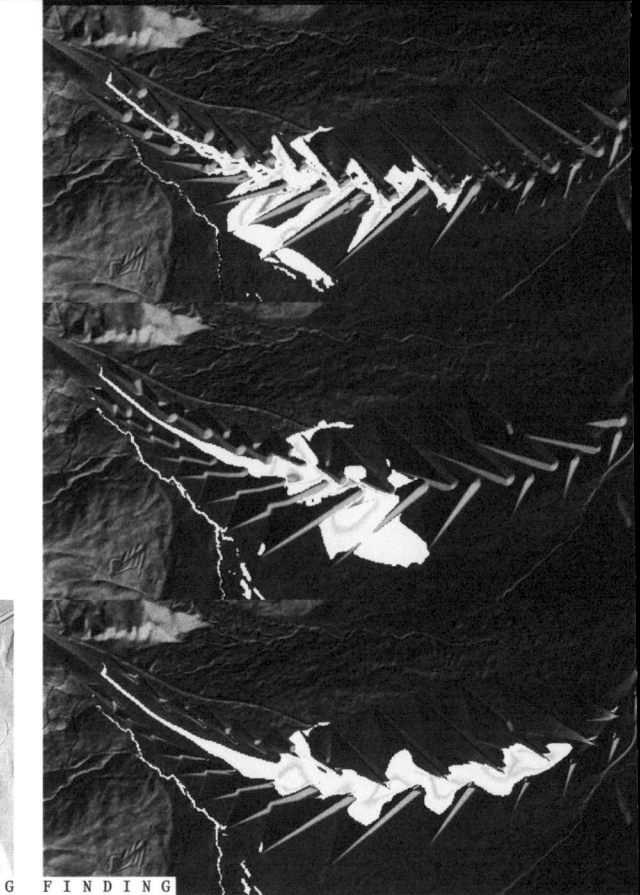

FORMING FINDING
FRAMING

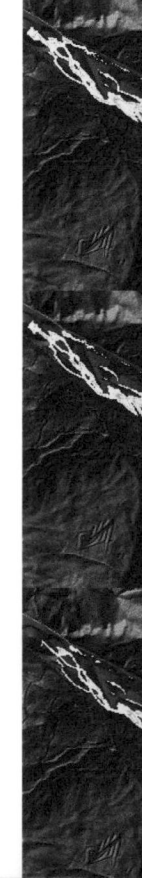

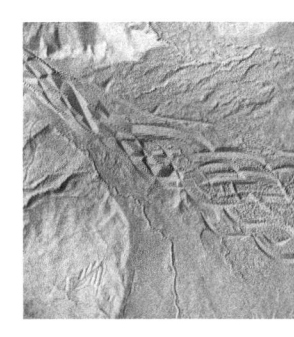

FRAMING
167
FINDING FORMING

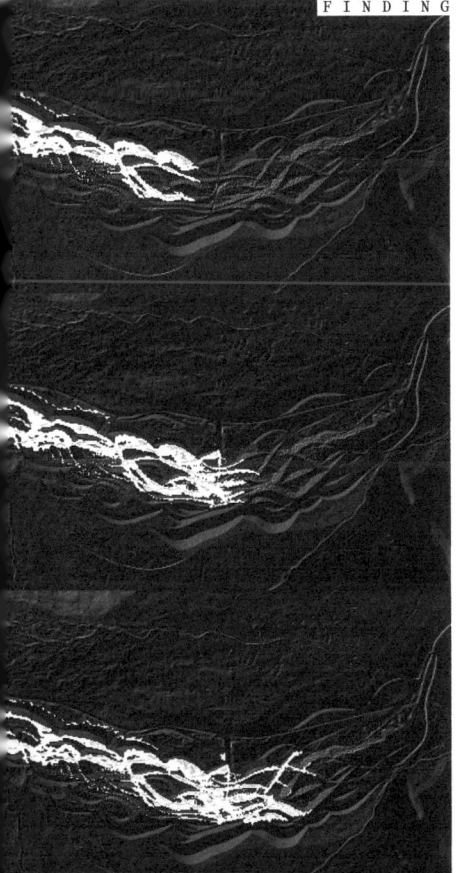

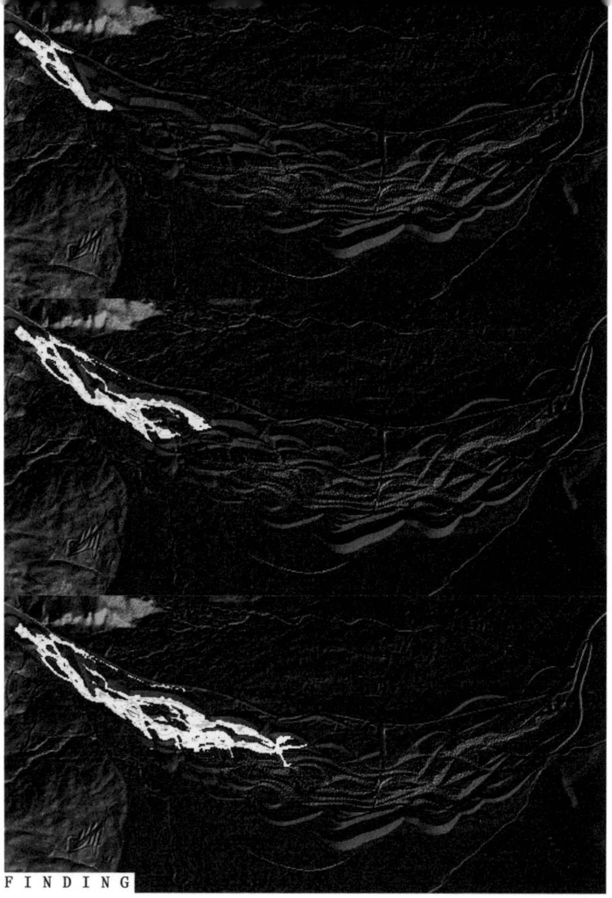

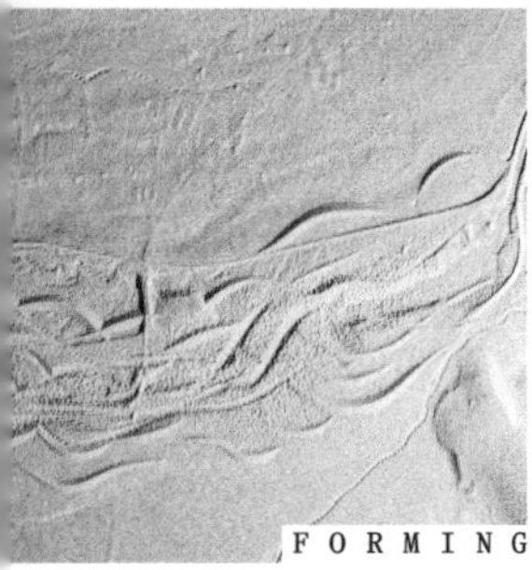

FORMING FINDING
168
FRAMING

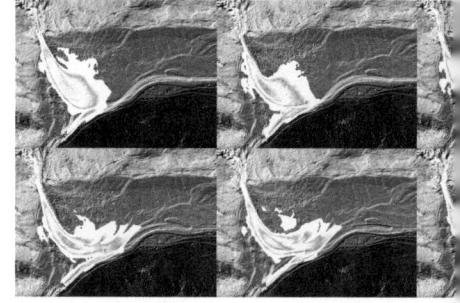

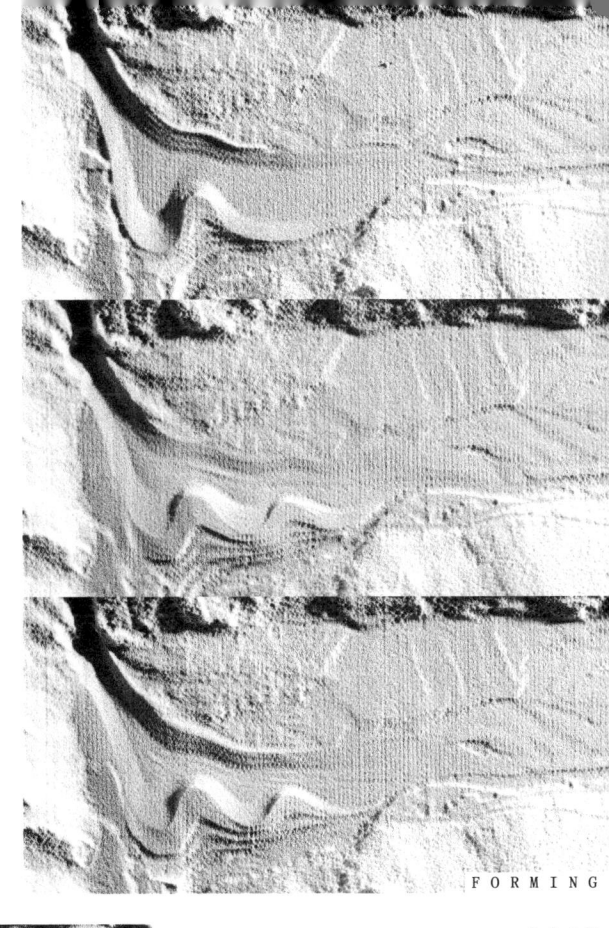

FRAMING

169

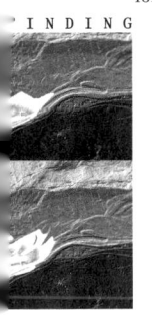
INDING FORMING
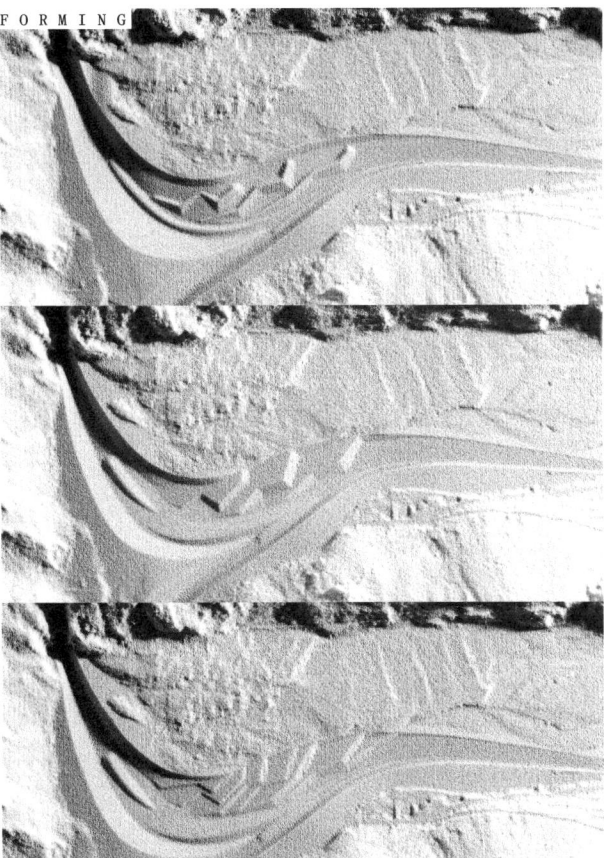

FORMING

FRAM
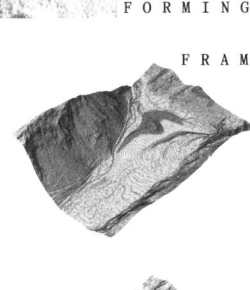

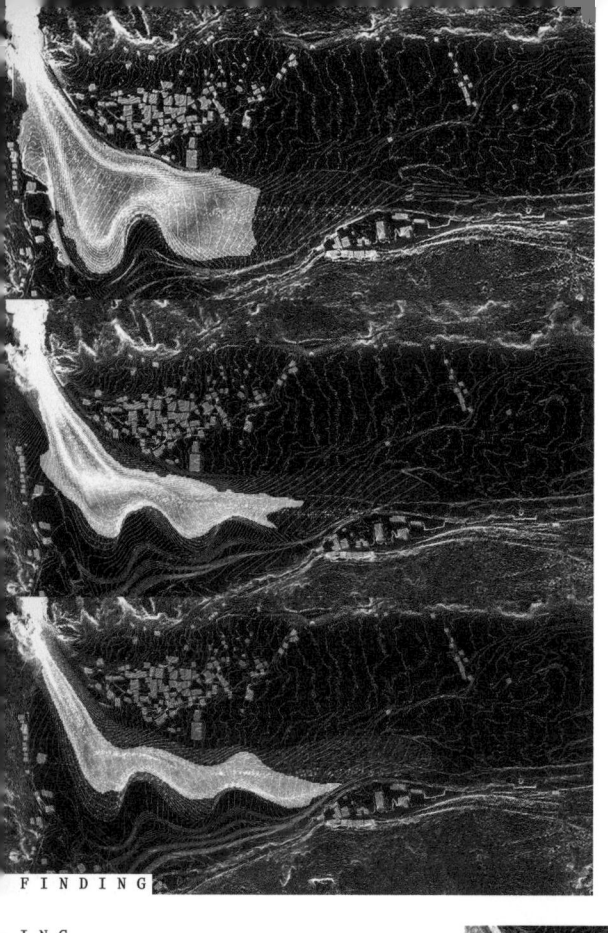

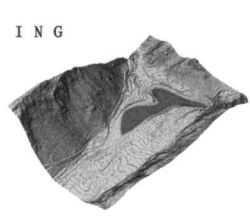

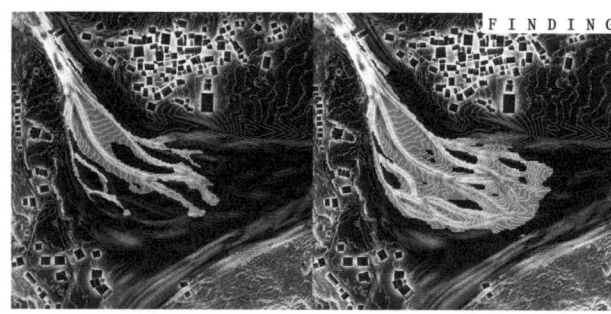

FINDING FRAMING

ING FINDING FORMING

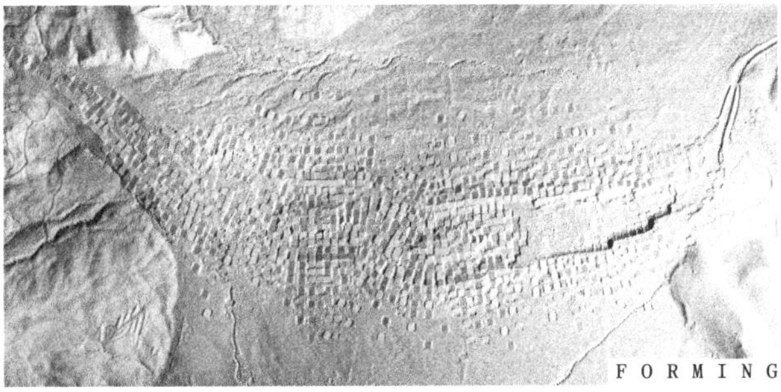

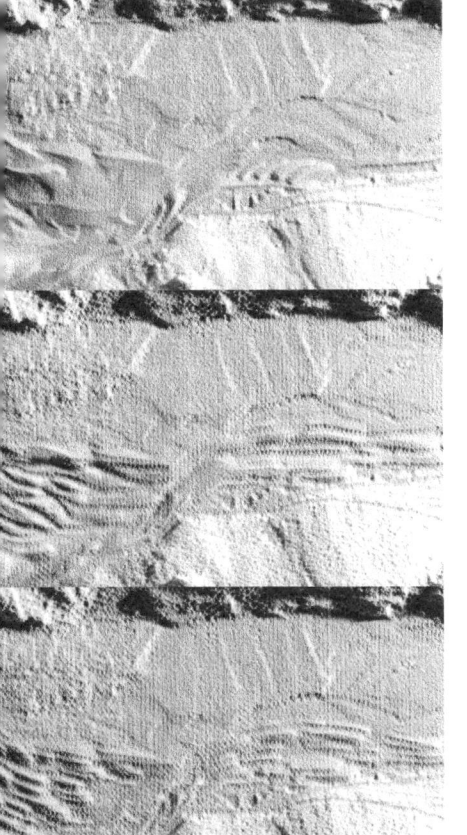

FORMING FINDING
172
FRAMING

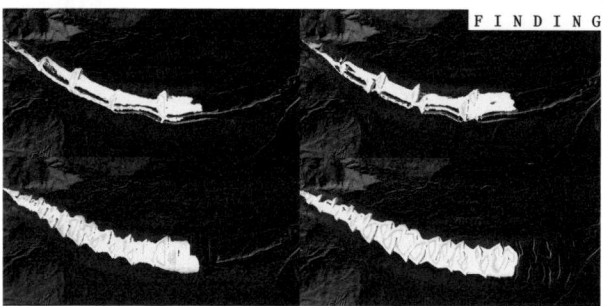

FRAMING
173
FINDING FORMING

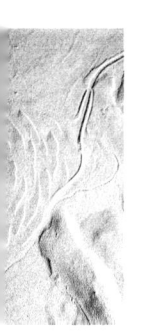

EVOLVING

"T0 is an original topography used as input data for the robotic process. T1 is a basin shaped by robotic processes that create accumulations according to the position of the boulders in T0.
T2 is the basin shaped by natural processes from the 400,000 cubic meters of debris flow material that covers T1 and fills up the collectors to create terraces.
T3 is the collaboration between the natural and robotic processes expressed by erosion / deposition and reconstruction.
This cycle shows that there can be constant change in the form of a topography."
KELLY MENG

"How much should I plan, and how much should I let it happen spontaneously? This was the very first question when considering the design of vegetation."
SAKIKO NODA

"'Living structures such as vegetation are deteriorated by exposure to debris flows. The ephemeral plants in the bottom of the basin are not the same as the perennial plants on the top, and it is not only the type of plant but also the varying levels of abundance that reveals a temporal dimension."
KELLY MENG

Over time soil development consumes energy and exports entropy through self-organization; a spontaneous order arises from local interactions between particles. Soil resilience relies on the ability to recover from shocks and disturbances to retain a state of equilibrium. This resilience is further linked to the concept of resurgence in all forms of life, demonstrating the ability of multiple species to recover on disturbed sites. Timely alignment of soil formation, plant formation, and robotic formation in mineral material systems facilitates a sustainable and resilient design approach.

By inputting energy and matter, a landscape emerges in space and time into a new material matrix with a distinct structural form. In leveraging entropic and resurgent processes, a new granular stratum forms under the influence of place, an environment, and the mechanized forces in a terrain. Instead of inert landscapes, regeneration and resurgence in a landscape's body become possible yet again.

Evolving is the act of guiding a process from the simple or rudimentary to the complex or advanced. The robotic earthmoving processes accomplish a newfound dynamic organization of the environment while creating a safe and purposeful landscape structure for ecologies of all kinds. The implications of such projects on existing habitats constitute, to a large extent, an effective instrument to achieve positive ecological connectivity, implying a greater use of the natural system's capacity for change. Over time, landscapes self-organize to create a new equilibrium with areas of enduring stability and others that continuously change.

LANDSCAPES ARE IN A CONSTANT STATE OF CHANGE. INVESTIGATIONS ARE MADE TO EXAMINE THE PREVAILING DYNAMICS THAT ALTER OR AFFECT TOPOGRAPHY OVER TIME. AUTONOMOUS ROBOTS MAINTAIN A PERFORMATIVE TOPOGRAPHY BY GENERATING A LANDSCAPE WITH FORMATION PROCESSES ALIGNED TO SEASONAL, HUMAN, OR NEWFOUND ROBOTIC TIMESCALES.

MATERIALIZING A GRADED LANDSCAPE FOUNDATION FROM THE SITE'S UNDERLYING FORM AND PROCESSES RESULTS IN A NEW MATERIAL STRATUM. THIS FOUNDATION ELICITS A BETTER UNDERSTANDING OF SYSTEMS, CYCLES, AND PROCESSES THAT IDENTIFY THE PARAMETERS NECESSARY TO INTERVENE WITH RULE-BASED DESIGN AND CONSTRUCTION PROCEDURES.

FORMING FINDING
179–191
FRAMING

STABLE EMBANKMENTS AND DIVERSE ECOLOGIES ARE GENERATED BETWEEN MOMENTS OF PERMANENCE AND DISTURBANCE USING THE FULL MINERAL AND BIOLOGICAL PALETTE OF THE SPECIFIC PLACE.

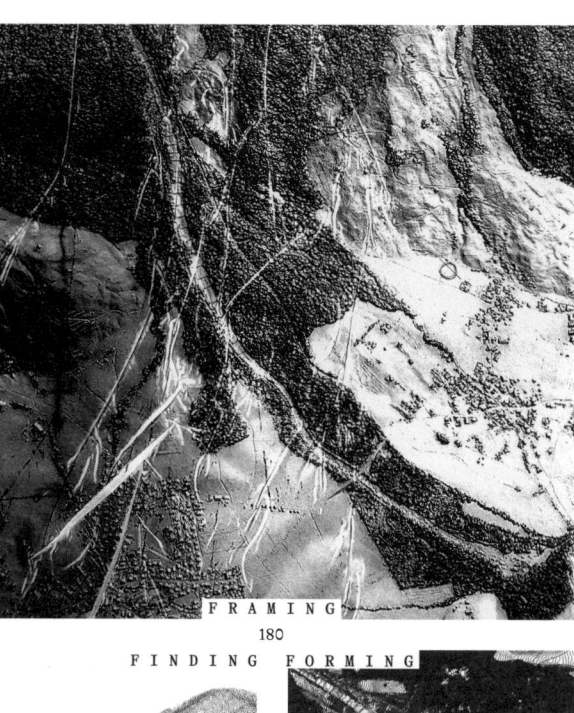

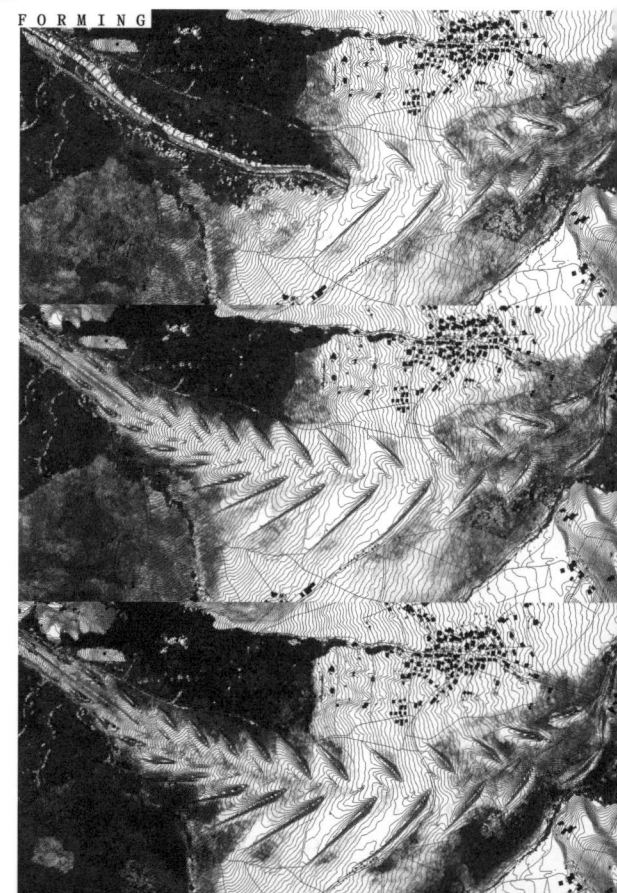

FRAMING
180
FINDING FORMING

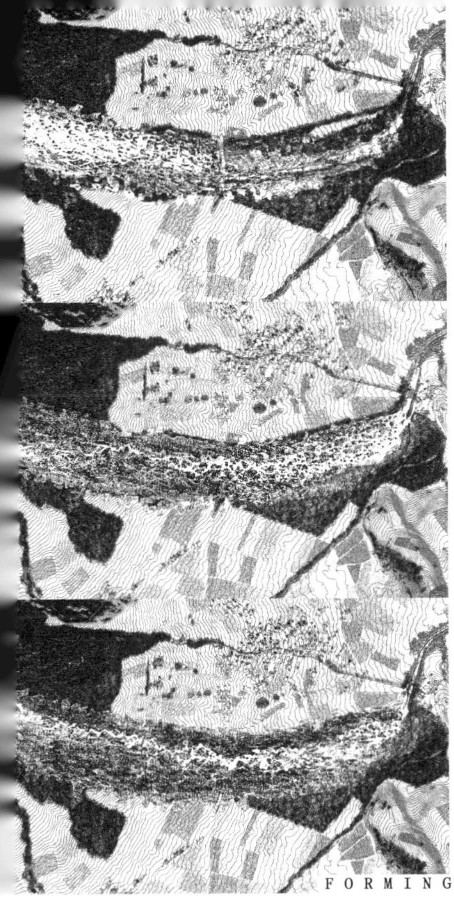

FORMING FINDING
FRAMING

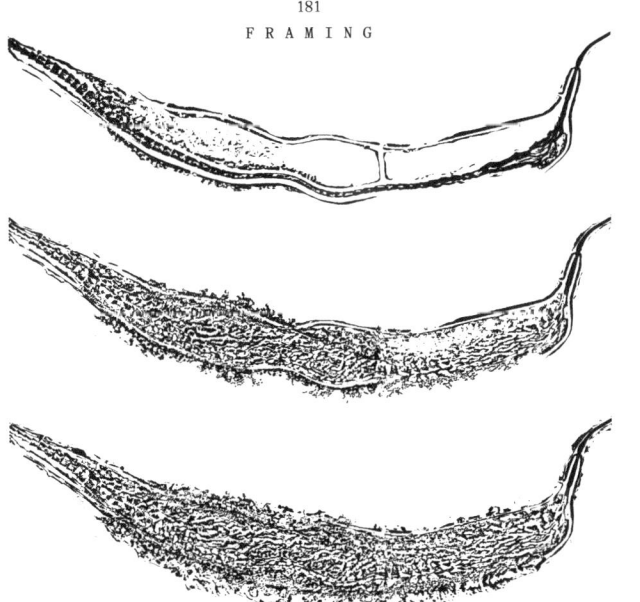

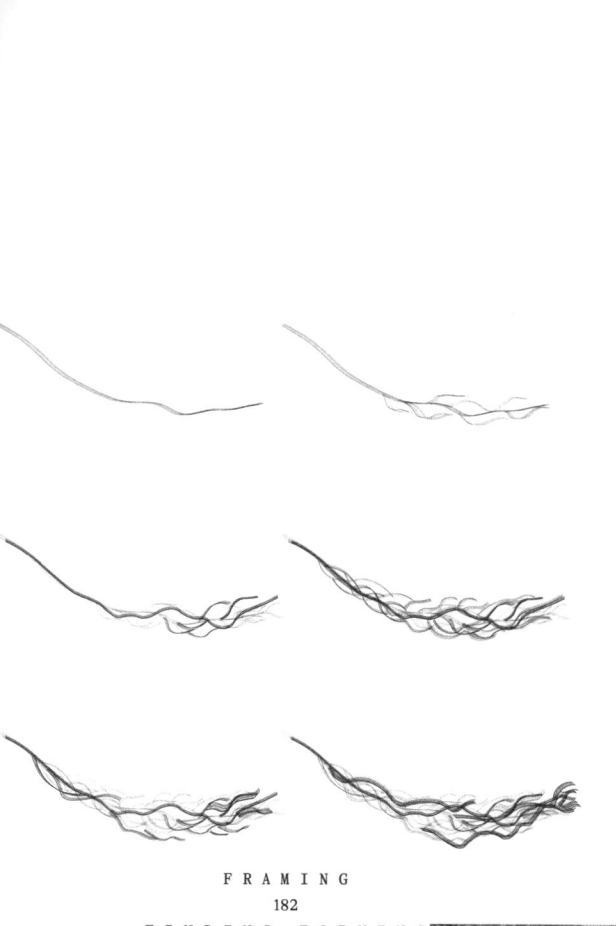

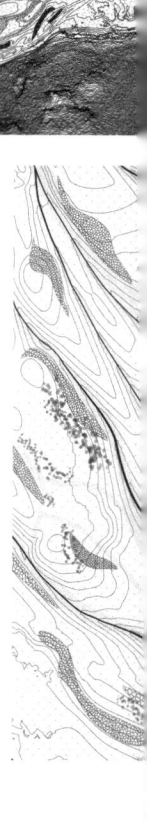

FRAMING
182
FINDING FORMING

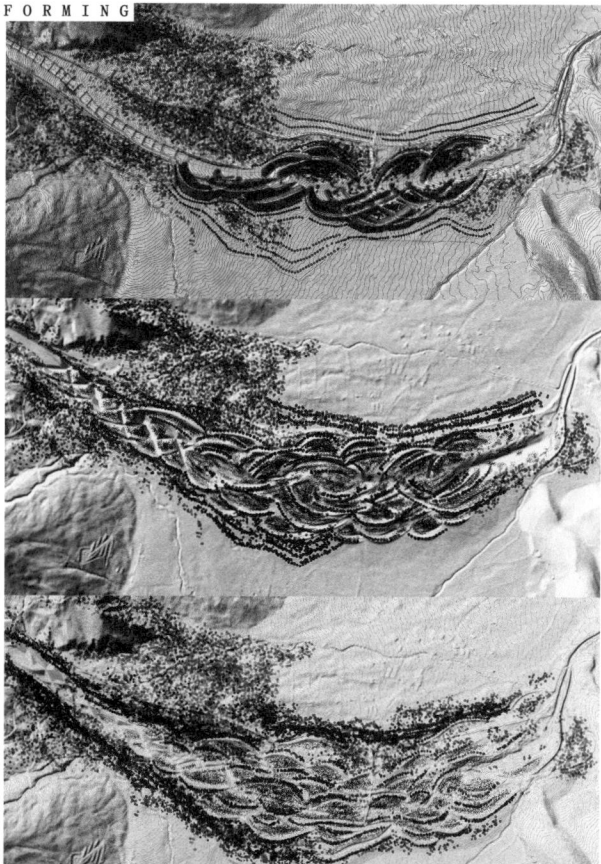

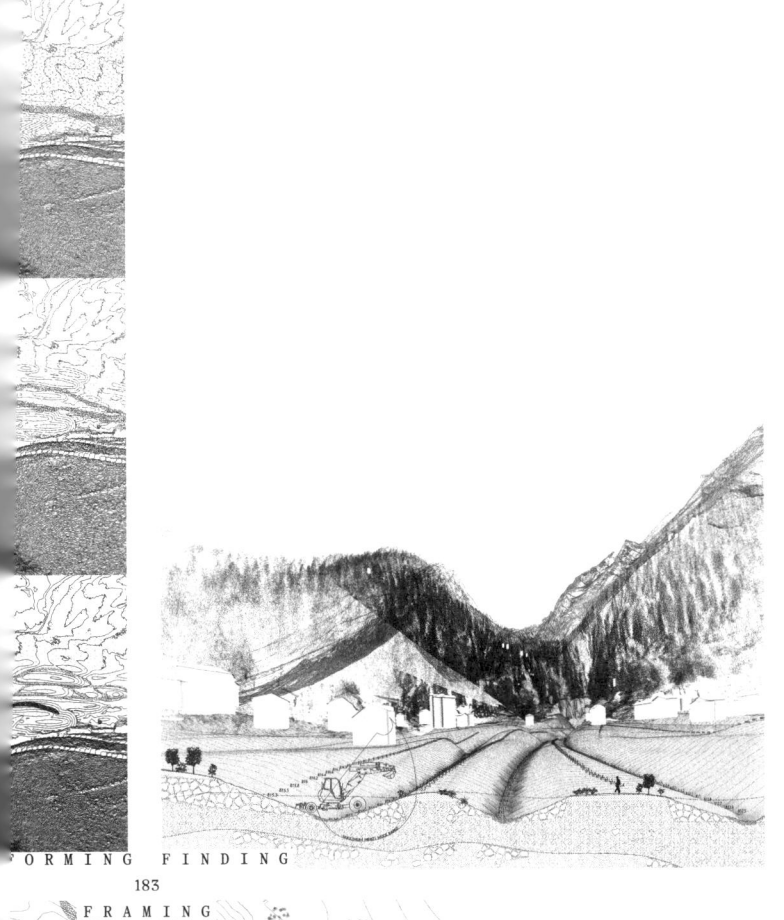
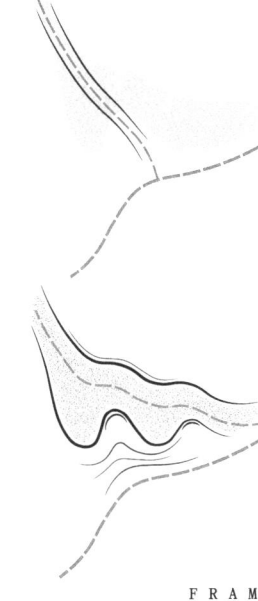

FORMING FINDING FRAM
183
FRAMING FINDING

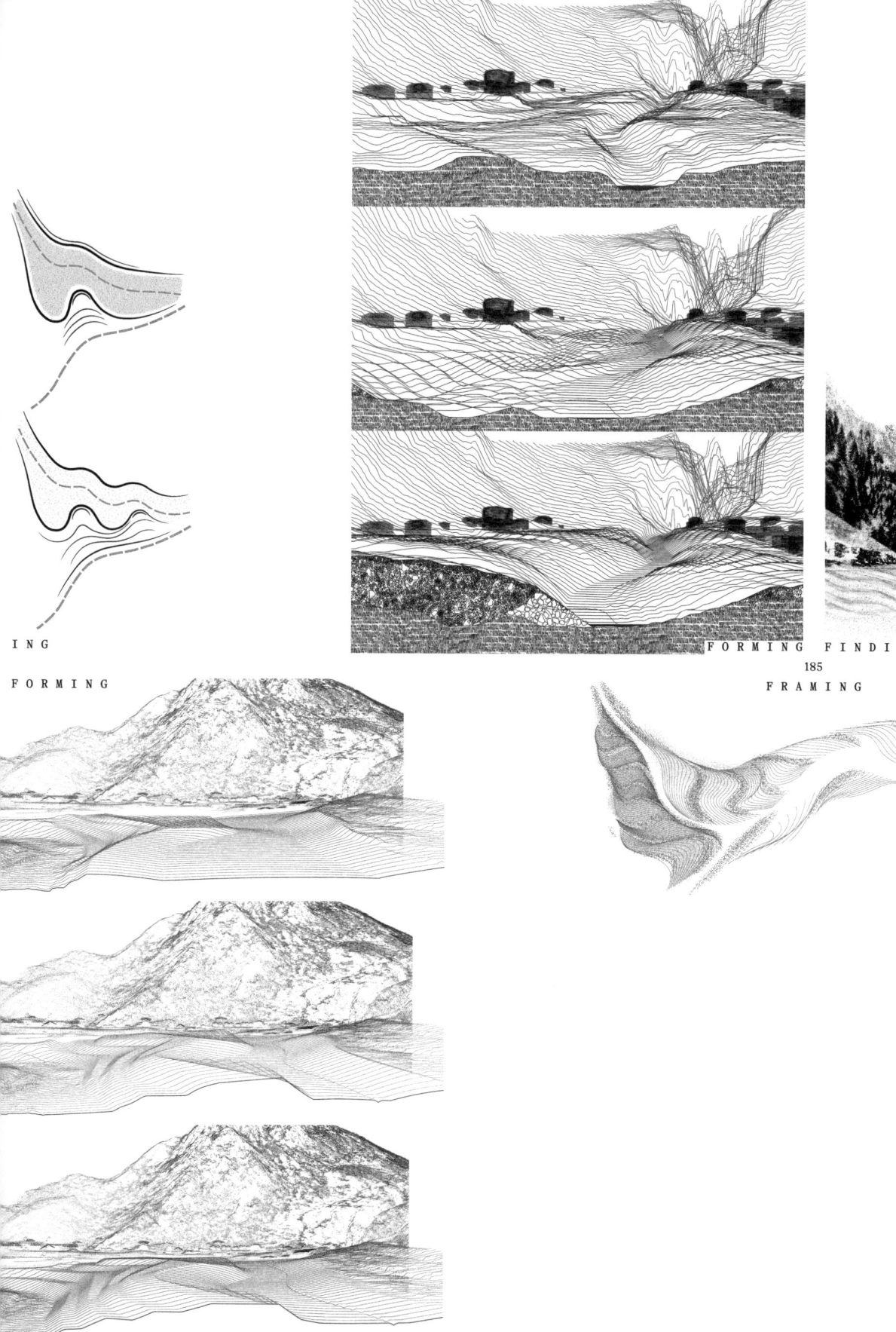

FRAMING
186
FINDING FORMING

FORMING FINDING
FRAMING

With the aid of multidisciplinary methods of digital surveying, computation, and robotic actuation, dynamic maintenance strategies can be conceived and constructed in a terrain. As the underlying form of the landscape continuously advances by natural processes, dynamic modeling and mobile manipulation strategies provide the necessary tools to maintain a site's dynamic equilibrium.

In anticipation of imminent earth movements, landscapes will forever be in the making. Instead of a finished appearance, the visual effect that a robotic excavator leaves behind is one of raw, unfinished imperfection. However, in this seemingly trivial effort against a sublime natural backdrop, the traces left behind by the robotic caretaker are deliberate, yet it may require the passage of time to fully appreciate its existence as part of the landscape.

Maintenance is the act that enables a condition to continue to be in existence. Responding to the challenges of the 21st century, a continually adapting maintenance strategy becomes a necessity. As such, robotic earthmoving procedures are orchestrated over long periods of time to ensure a resilient landscape future.

"The robotic process was split into two tasks—implementation and maintenance. While the design of the system is simple, the process becomes far more interesting when you begin to think about its maintenance."
CASPAR TRUEB

"We have to closely study the robotic dimensions of HEAP so as to understand how to design the landscape around it. It is an interdependent relationship where the landscape is shaped by HEAP, but in a manner that accommodates the perpetual construction of the autonomous excavator."
LEE LIP JIANG

"To redistribute a 200,000 cubic meter debris flow, we calculated that it would take a Menzi Muck M454 with 55 kilowatts of power working continuously for 202 days. Considering the travel distance and time required, the time needed would extend to about a year."
HANNAH KILIAN

MAINTAINING

THE SPECIFIC DIMENSIONS OF THE ROBOTIC PLATFORM
DETERMINE THE EXTENT OF A LANDSCAPE'S TRANSFORMATION
AS CHOREOGRAPHED BY THE DESIGNER, GIVING SHAPE TO
A NEW ROBOTIC SCALE.

FRAMING

FINDING FORMING

THE ROBOTIC PRESENCE IN A LANDSCAPE ESTABLISHES
NOVEL LANDFORMS AND A RENEWED UNDERSTANDING
OF ONGOING PROCESSES BETWEEN SHORT-TERM CHALLENGES
AND LONG-TERM VISIONS.

ANTICIPATING FUTURE LANDFORM PROFILES THAT EXPAND
AND CONTRACT THROUGH TIME. THE EVOLVING
LANDSCAPE REVEALS EVIDENCE OF THE INTERACTION
BETWEEN TECTONIC, TOPOGRAPHIC, CLIMATIC, AND ROBOTIC
PROCESSES AFFECTING THE LANDSCAPE AS A WHOLE.

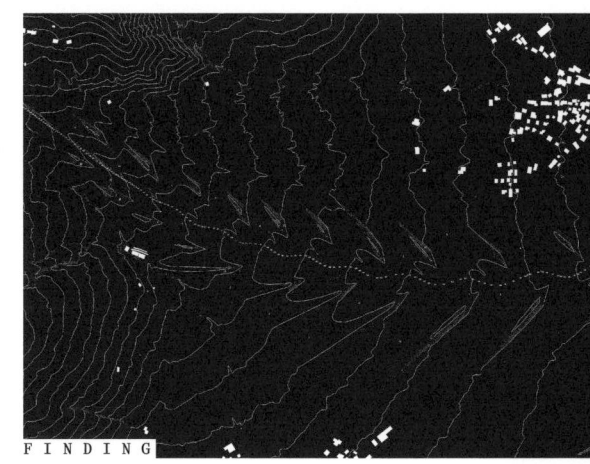

FORMING FINDING
194
FRAMING

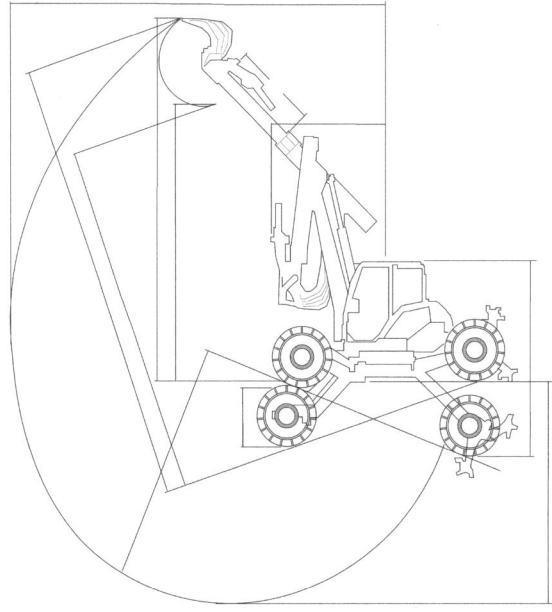

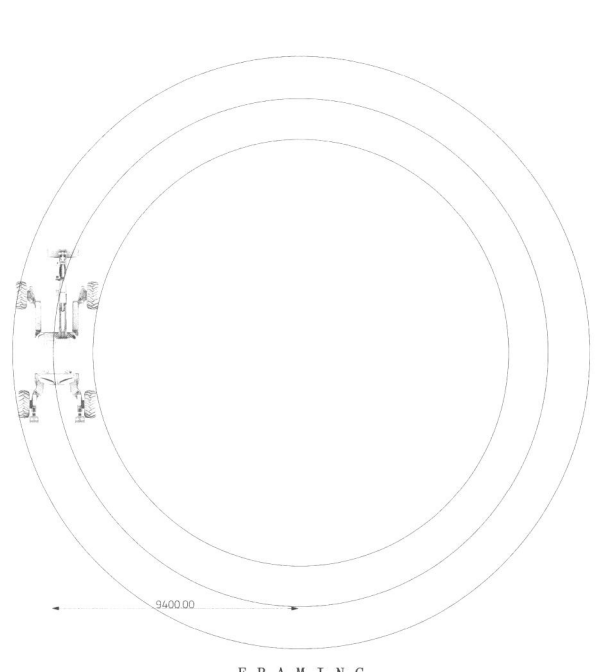
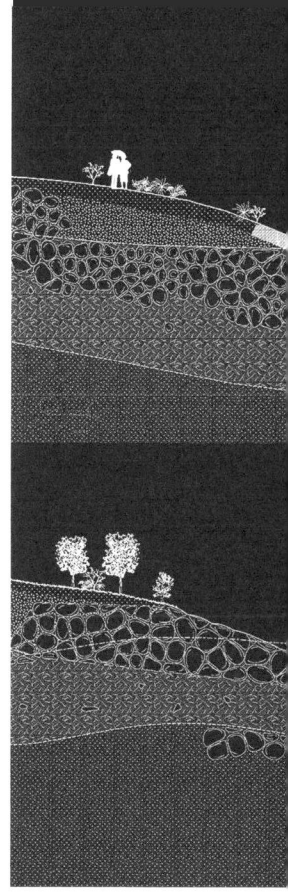

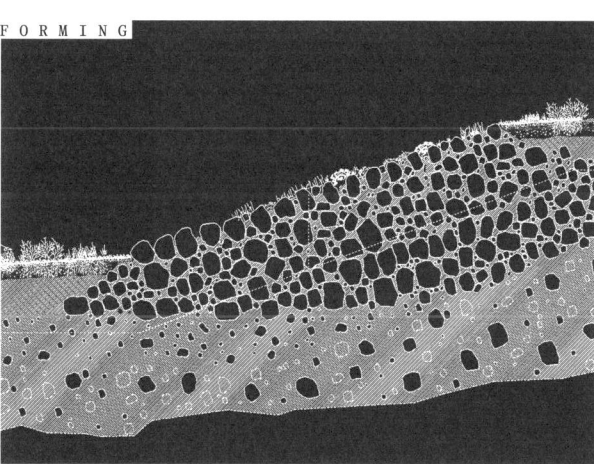

FRAMING
195
FINDING FORMING

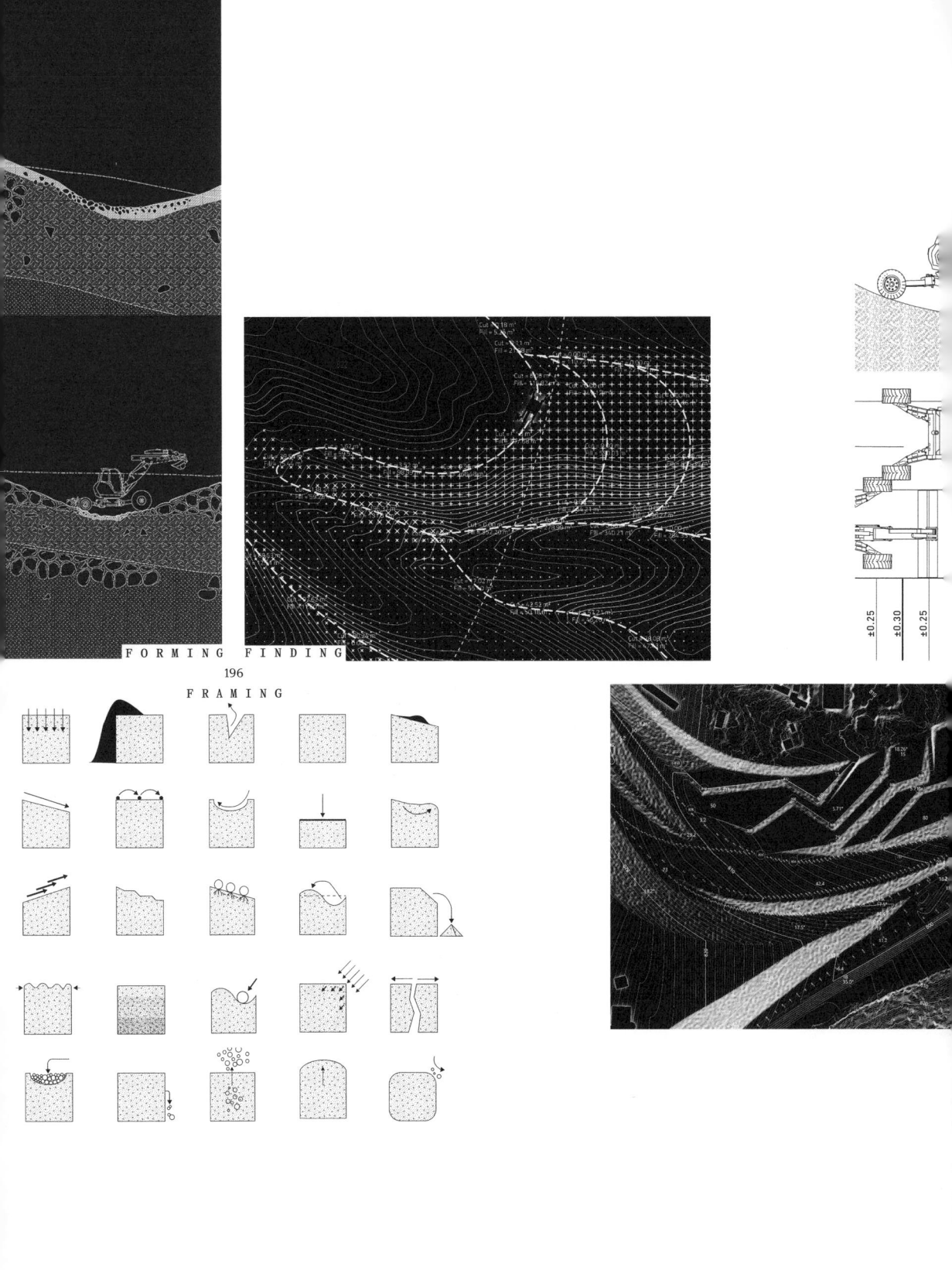

FORMING FINDING
FRAMING

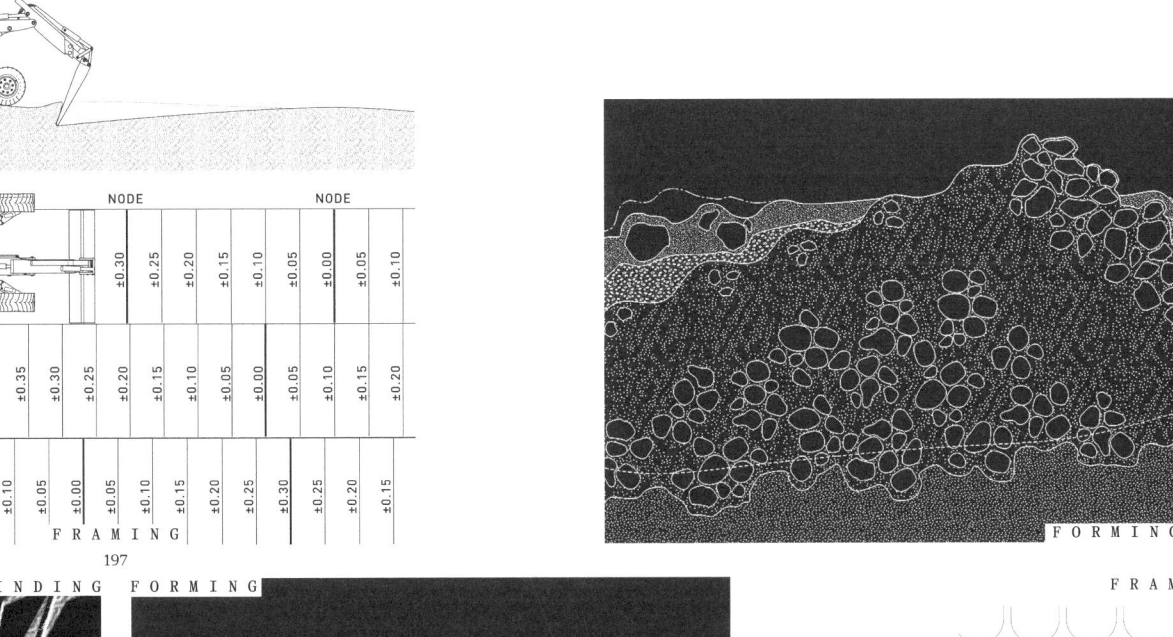

FORMING

FRAM

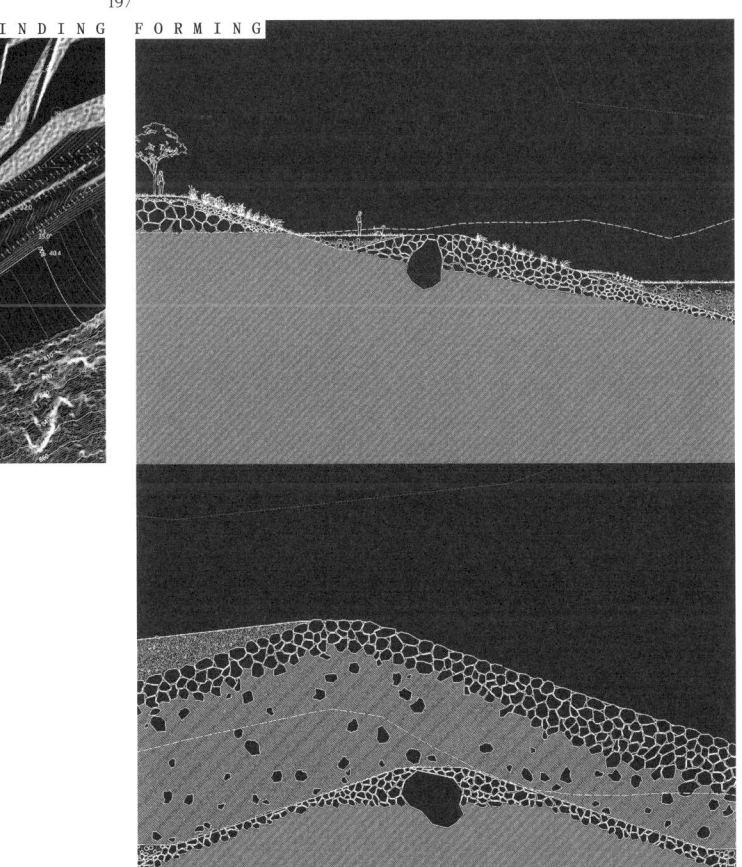

INDING FORMING

FRAMING

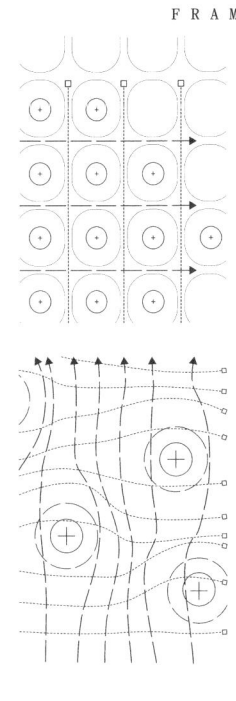

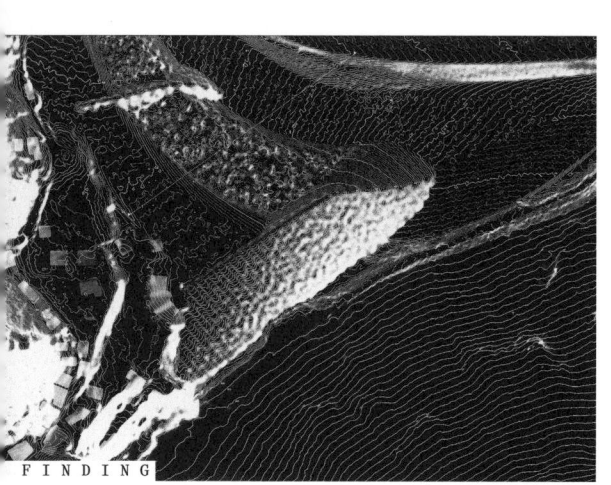
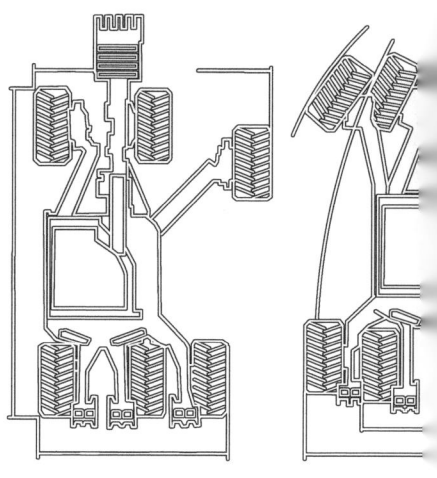
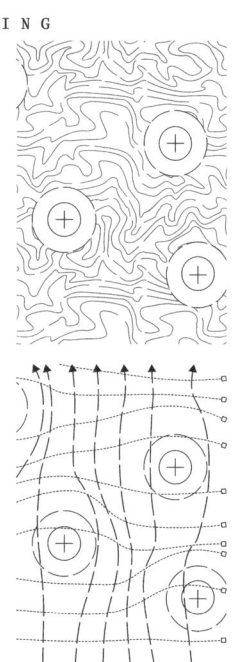

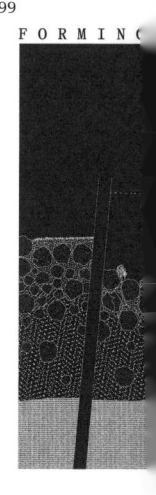

FINDING   FRAMING

ING   FINDING FORMING

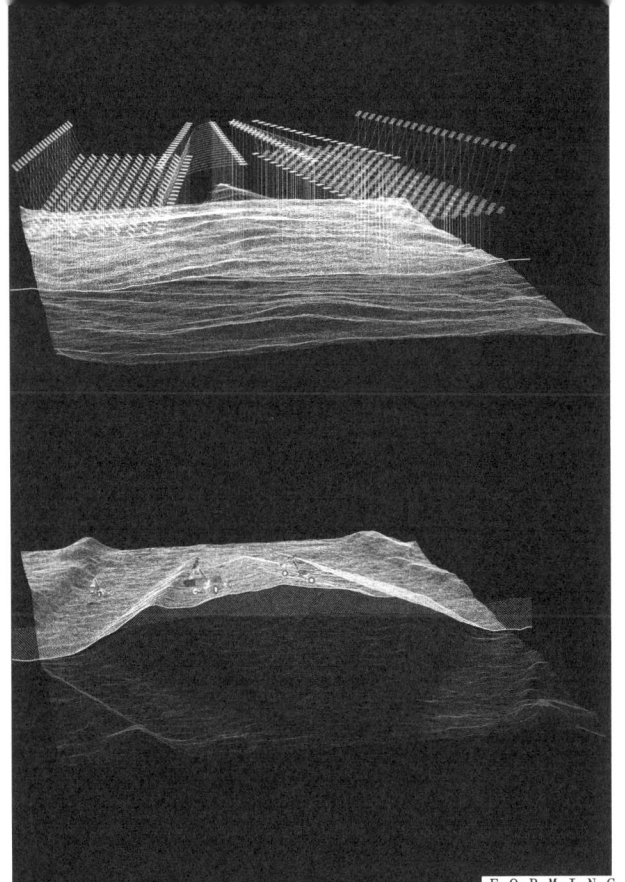
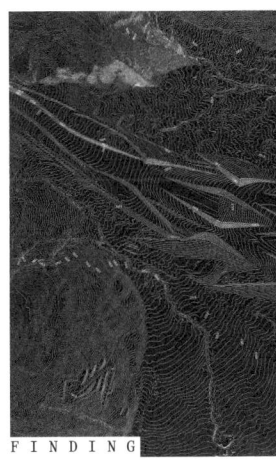
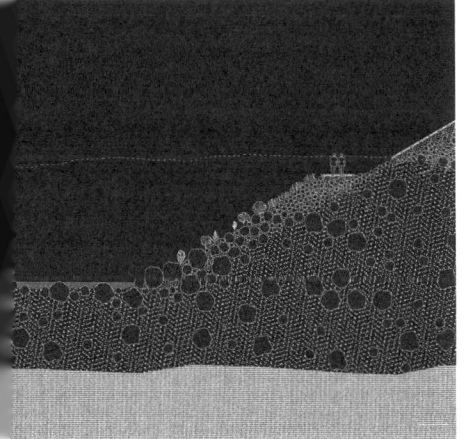

FORMING FINDING
200
FRAMING

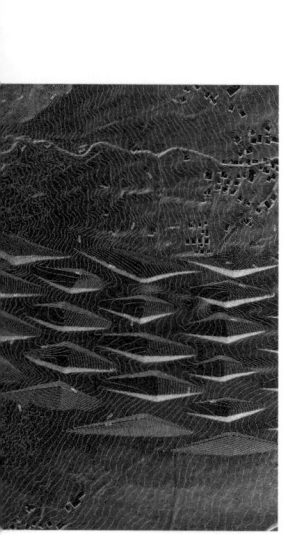
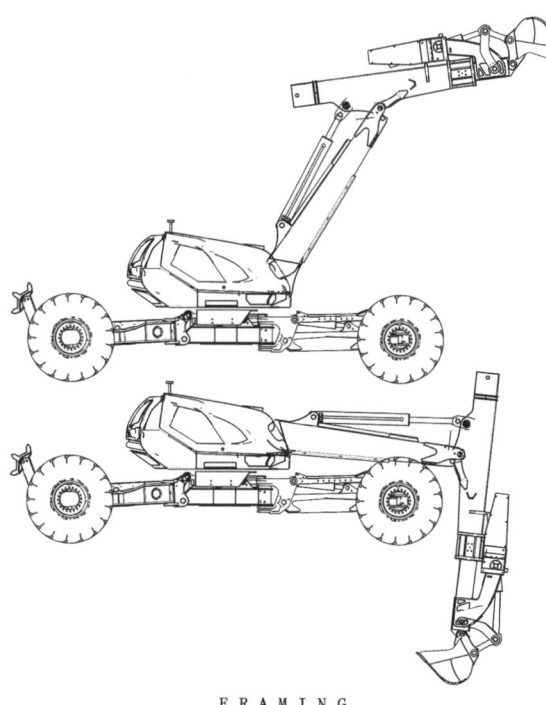

FRAMING
201
FINDING FORMING

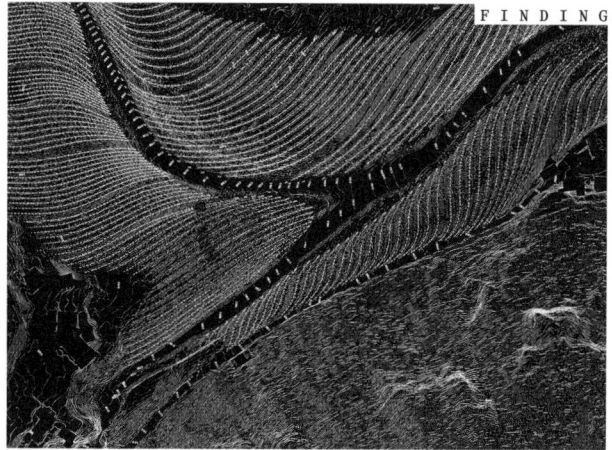
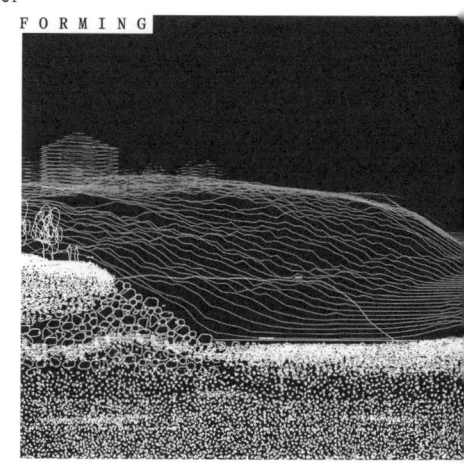

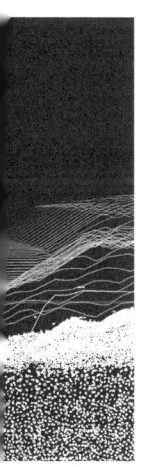

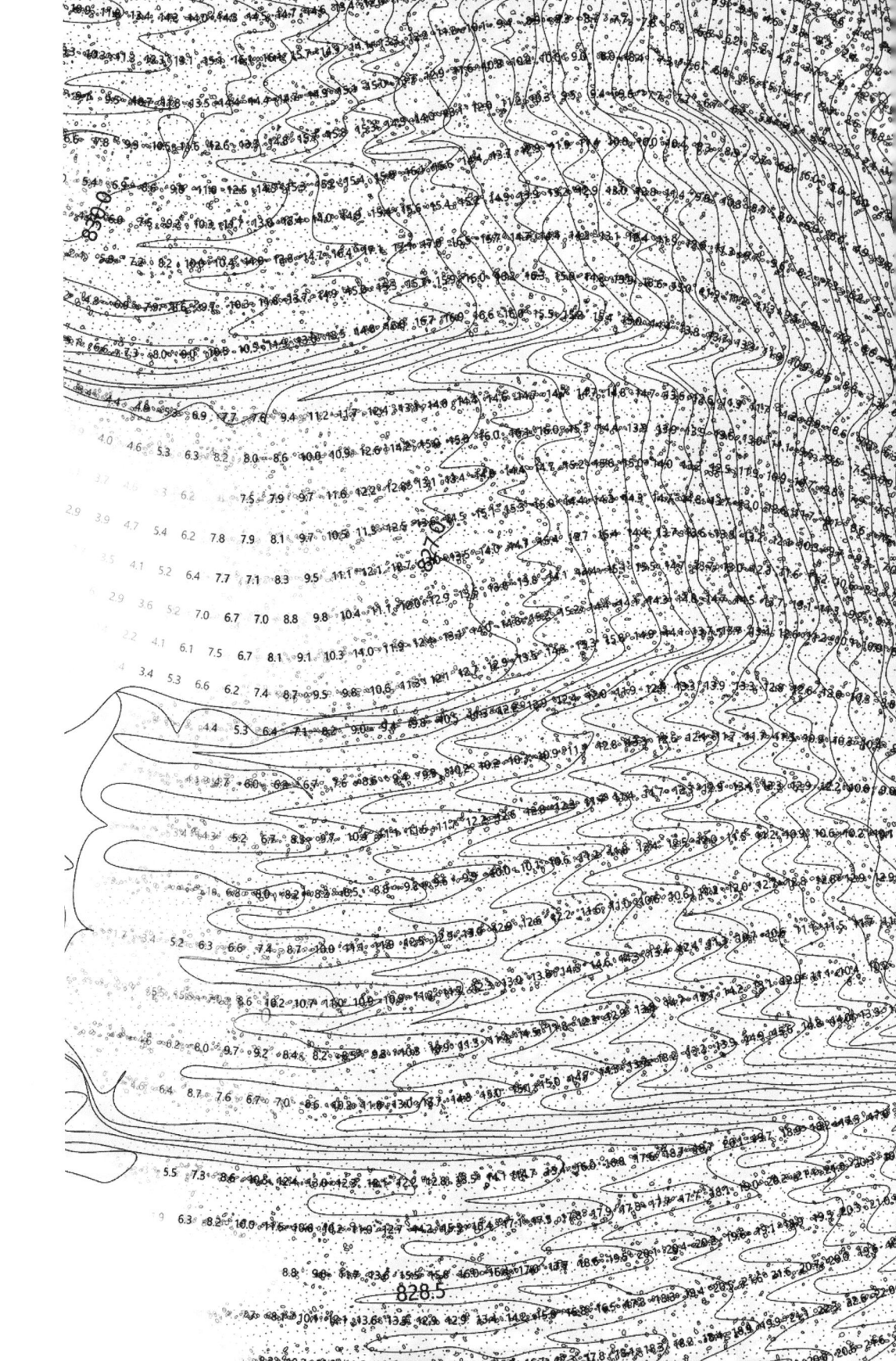

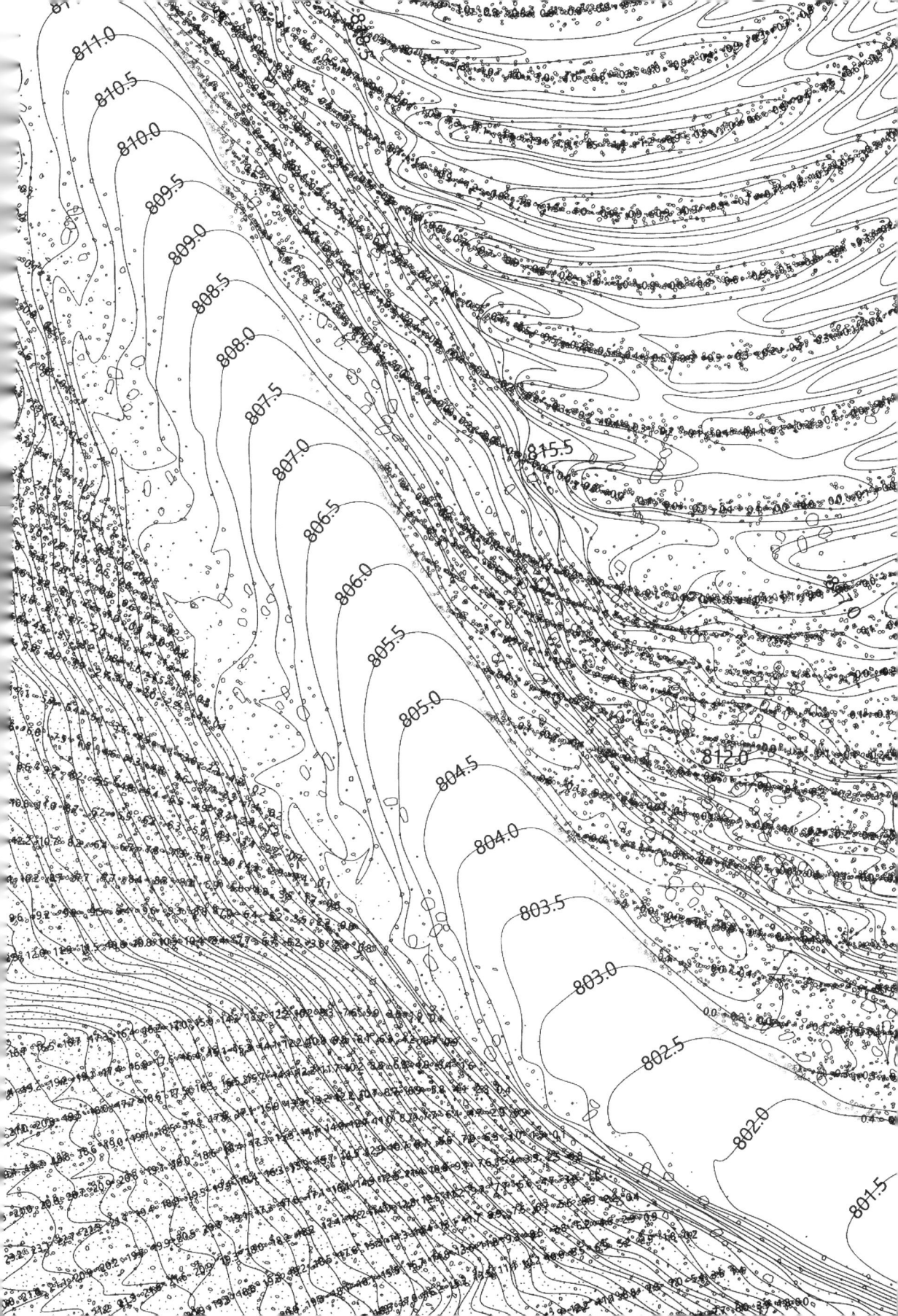

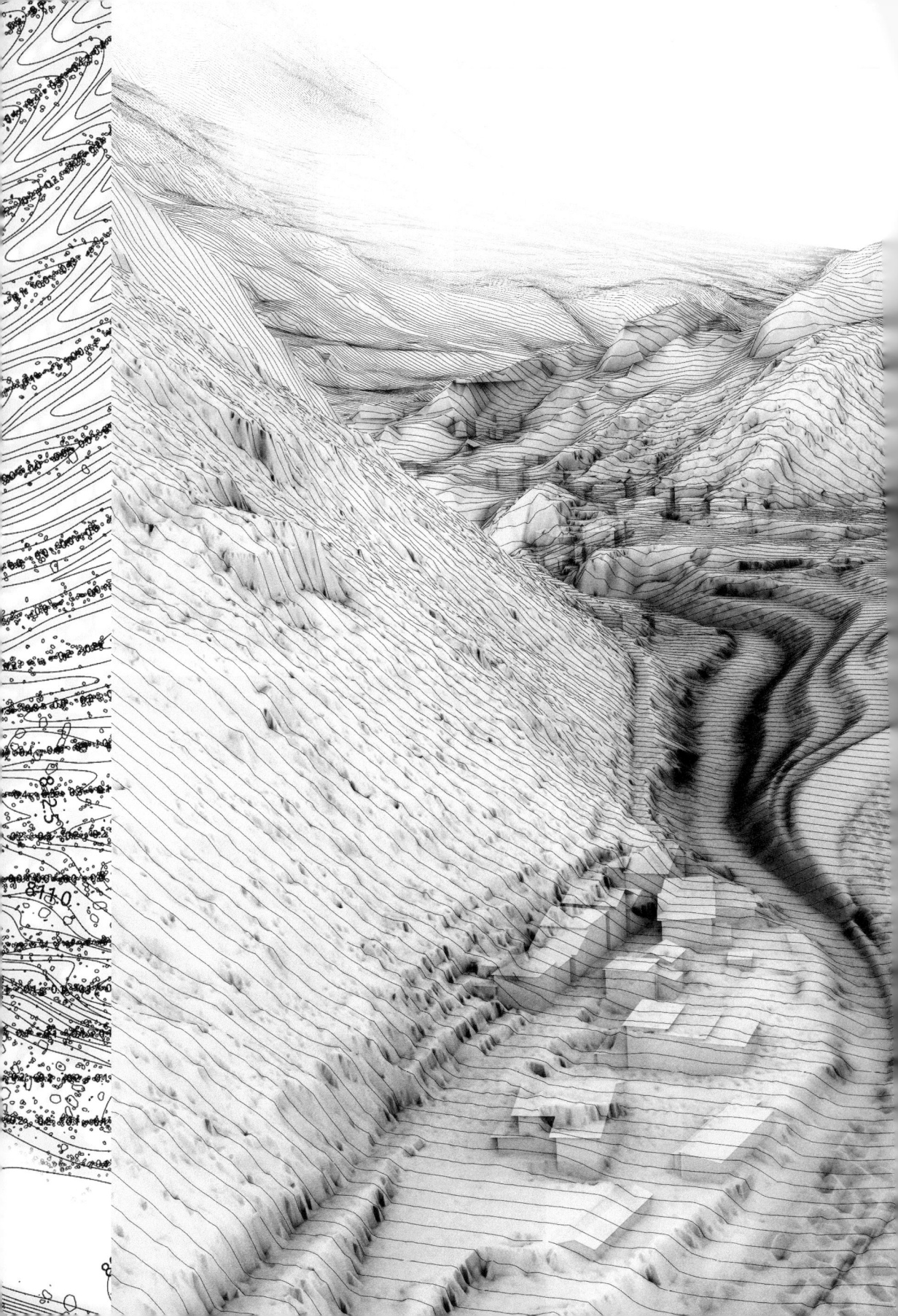

IMPRINT

CONCEPT
Ilmar Hurkxkens, Fujan Fahmi
COPY EDITING
Jack Eden, Rotterdam
PROOFREADING
Colette Forder, Marseille
DESIGN
Janic Fotsch and Pascal Sennhauser, Zurich
LITHOGRAPHY
PRINTING
AND BINDING
DZA Druckerei zu Altenburg GmbH, Thuringia
TYPEFACE
GT Alpina, Grilli Type
PAPER
Metapaper Extrarough white

All texts are the copyright of the authors

Geospatial data is copyright of swisstopo

All images are subject to copyright of collaborators and students under the Chair of Landscape Architecture, Professor Christophe Girot, and Gramazio Kohler Research, ETH Zurich, except p.67 © Dominic Jud, pp.70, 73 © Mathias Bernhard

© 2021 ETH Zurich, Chair of Landscape Architecture of Professor Christophe Girot, Gramazio Kohler Research, and Park Books AG, Zurich

Park Books
Niederdorfstrasse 54
8001 Zurich
Switzerland
www.park-books.com

This research was supported by the NCCR Digital Fabrication, funded by the Swiss National Science Foundation (NCCR Digital Fabrication Agreement #51NF40-141853). Park Books is being supported by the Federal Office of Culture with a general subsidy for the years 2021–2024.

All rights reserved; no part of this publication may be reproduced, stored in a retrieval system, or transmitted in any form or by any means electronic, mechanical, photocopying, recording, or otherwise, without the prior written consent of the publisher.

ISBN 978-3-03860-254-5

ETH zürich

D**ARCH**

Prof. CHRISTOPHE
GIROT

GRAMAZIO
KOHLER
RESEARCH

RSL Robotic Systems Lab

National Centre of Com in Research Digital Fabrication